The One Culture?

The One Culture?

A CONVERSATION ABOUT SCIENCE

Edited by Jay A. Labinger and Harry Collins

The University of Chicago Press
Chicago and London

Jay A. Labinger is a research chemist and administrator of the Beckman Institute at the California Institute of Technology.

Harry Collins is Distinguished Research Professor of Sociology and director of the Centre for the Study of Knowledge, Expertise, and Science (KES) at Cardiff University.

The University of Chicago Press, Chicago 60637
The University of Chicago Press, Ltd., London

Printed in the United States of America
10 09 08 07 06 05 04 03 02 01 1 2 3 4 5

ISBN: 0-226-46722-8 (cloth)
ISBN: 0-226-46723-6 (paper)

Library of Congress Cataloging-in-Publication Data

The one culture? : a conversation about science / edited by Jay A. Labinger and Harry Collins.
p. cm.
Includes bibliographical references and index.
ISBN 0-226-46722-8 (cloth) — ISBN 0-226-46723-6 (paper)
1. Science—Social aspects. 2. Science and state. 3. Science—Philosophy. I. Labinger, Jay A. II. Collins, H. M. (Harry M.), 1943– III. Title.
Q175.55 .O54 2001
303.48′3—dc21

00-013097

♾ The paper used in this publication meets the minimum requirements of the American National Standard for Information Sciences—Permanence of Paper for Printed Library Materials, ANSI Z39.48-1992.

Contents

Part Two Commentaries

Part Three Rebuttals

Preface

While being an academic may not be the best paid or most prestigious job in the world, it does have many compensations. Prominent among these is the opportunity to make discoveries or contribute to the deep debates of the age and to work in an environment characterized by honesty, integrity, and careful, high quality thought. For some of us working in the borderlands between the natural sciences and the humanities and social sciences, however, that environment seemed to be degenerating in the early 1990s with the outbreak of a particularly nasty quarrel known as the "science wars." In these arguments—between those who favored a traditional, "hard-edged" model of the natural sciences and certain social scientists and humanities scholars who wanted to look at science in new ways—the quest for quick public victories took priority over scholarship, and the quality of debate fell disastrously low.

Fortunately, even in the middle of the war, small groups of scientists and social scientists continued disagreeing in something like the old way. One of the editors of this volume—sociologist Collins—found himself having discussions of this more recognizable kind with the other—chemist Labinger—as well as with other scientist critics, including physicists Peter Saulson and David Mermin. Labinger had written a critique of science studies in *Social Studies of Science,* the leading journal of the science studies field; Saulson, a subject in one of Collins's case studies, had written Collins a letter describing his misgivings after reading one of Collins's books; and Mermin (who is at the same university as Collins's coauthor, Trevor Pinch) had written a fiercely critical review of their book *The Golem* in the journal *Science.* Crucially, all these commentaries started by outlining what they saw as positive about the work they were attacking and engaged seriously with the ideas before launching the attack; there was a sharp sword in one hand but an olive branch in the other. Criticism, fierce though it was, still resembled traditional academic argument rather than cheap point scoring. These dialogues held out the hope for better and more productive times.

In May 1997 Mike Nauenberg, a physicist at the University of California at Santa Cruz, organized a small conference which brought together some of the critics and the criticized. There, Collins was able to meet Alan Sokal and to continue his discussions with David Mermin. Interestingly, in the setting of that conference, Mermin and Collins often found themselves pushed together by common disagreement with Sokal. Around the same time, Collins, then at Southampton University, was organizing the so-called Southampton Peace Workshop, which took place in July 1997. Sokal was unable to come, but the participants included Labinger, Mermin, and Pinch, as well as several other scholars representing diverse fields: physics, history of science, literary theory. For two days, one of which was spent cruising round Southampton water in a small motorboat, the eight conferees were closeted together for intense discussions. Only on the third day, after some mutual trust and understanding had been built up, did the workshop become a public event. This contrasted with other science wars forums, where opposition was usually absent and it often seemed that a public display of scorn was the chief purpose. It was after the Southampton workshop that the idea for the present volume was born.

Working on this book has been an (almost) unalloyed pleasure. Some participants initially felt they might not be able to offer much but were drawn in as the successive rounds unfolded. Nearly everyone kept to the timetable and provided manuscripts that left us relatively little to do.[1] There have been some wonderful and novel surprises mingled with the more predictable positions. And perhaps most surprisingly of all, the editors saw "eye-to-eye" pretty well all the way through.[2]

We sent the introduction and conclusion out to authors before finalizing them. Most responded with helpful ideas, for all of which we are grateful. We owe a particular debt to David Mermin, who sent us many pages of notes suggesting stylistic and substantive adjustments, nearly all of which we incorporated. Alan Sokal and Jean Bricmont also went well beyond the call of duty and helped us to round off the book with a lively discussion of methodological relativism, which is an invitation to further debate.

1. One of the publisher's manuscript reviewers congratulated the editors on keeping all the authors up to scratch in their writing styles. While we would love to accept the accolade, we actually did very little in that regard; the articles, with rare exceptions, are just what the authors sent us.

2. The most contentious issue—whose name should go first?—was settled by a coin toss, after a more sophisticated decision-making technique (an e-mail version of rock, paper, scissors) somehow ended in a perpetual stalemate.

Finally, our editor at the University of Chicago Press, Susan Abrams, has provided suggestions and encouragement throughout while guiding us through the technicalities and pitfalls of multiple authorship. We thank everyone involved while accepting the blame for everything that's still wrong.

Jay A. Labinger
Harry Collins

Chapter 1

INTRODUCTION

THE "SCIENCE WARS" AND THE "TWO CULTURES"

In the spring of 1996, a heated but obscure academic debate entered the public arena. The immediate cause was Alan Sokal, a physicist from New York University. He wrote an article in a special issue, entitled "The Science Wars," of the cultural studies journal *Social Text* (1996b). Sokal's piece appeared to be an analysis of certain aspects of physics and mathematics, conducted from the viewpoint of what has come to be known as postmodernism. It seemed to argue that language, politics, and interests, rather than objective reality, determine the nature of scientific knowledge, and it was written in the characteristic style of postmodernist discussions of science—pedantic bibliographic referencing, extravagant praise for an elite group of thinkers, and an elaborate language which some might call "jargon." After publication Sokal immediately issued a disclaimer in the journal *Lingua Franca* (1996a), revealing that the first article was a parody, intended as an experiment to determine whether the cultural studies community, and the journal's editors in particular, could tell the difference between serious scholarship and deliberate nonsense.

This well-crafted and nicely executed hoax embarrassed some trendy cultural critics and caught the fancy of the media. Sokal's piece attracted a host of new participants to the argument. Some of them (not represented in this book) felt empowered to comment on the stupidity, venality, and general worthlessness of any nonscientist who proffered an opinion on science, or on the naïveté and arrogance of scientists who ventured beyond the confines of their expertise. For this subset, scholarly study, deep familiarity with the debate, and close analysis of the arguments were not always considered prerequisites for entry into the fray. Indeed, over the next few months, most of the debate took place in the popular media—daily newspapers and weekly newsmagazines.

This, then, was the high point (or low point) of the science wars. The wars had actually begun quietly several years earlier. In 1992 a pair of books by American physicist Steven Weinberg and British biologist Lewis Wolpert included attacks on the new schools of science criticism, but those constituted only small chapters in their volumes and received little attention. In 1994, accounts of a debate between Wolpert and Harry Collins and of a meeting of the US National Association of Scholars (prominently featured in the *Times Higher Education Supplement* and the *Boston Globe,* respectively) brought about the first stirrings of awareness outside the academic community and helped set the intemperate style of what was to come.

Following these skirmishes, the 1994 book *Higher Superstition: The Academic Left and Its Quarrels with Science,* by American biologist Paul Gross and American mathematician Norman Levitt, can be identified as the formal declaration of (science) war. Gross and Levitt examined literary, sociological, and feminist approaches to science studies and topics such as radical environmentalism, AIDS patient activism, animal rights activism, and Afrocentric history. They argued that these seemingly disparate areas in fact have several things in common: intellectual laxity above all, but also hostility to science and a political agenda summed up by the sobriquet "Academic Left." An organized response to this challenge appeared as a collection of articles published in the Science Wars issue of *Social Text,* which included Sokal's "Trojan horse" and triggered the aforementioned explosion.

What is under debate, and why should it be of much (if any) concern to the vast majority of scientists and humanists as they go about their business? The origins of the argument go back at least to Plato, but a convenient starting point is C. P. Snow's much discussed "The Two Cultures," given as the Rede Lecture at Cambridge in 1959 and later published. Snow decried the sharp separation between the scientific culture on one hand and the literary/humanistic culture on the other. His strongest objections, perhaps, were to the asymmetry he perceived: for a scientist to be unfamiliar with the literary canon was to be considered uncultured by the establishment, whereas the converse was accepted virtually as a matter of course. But his central point was that the two groups that between them are responsible for most of mankind's understanding of itself and the world it inhabits do not know how to talk to each other.

This was not a new observation; similar complaints about the invisibility of science in liberal education had been raised by T. H. Huxley (1900) and Henry Adams (1918). It was also, of course, not entirely true—there have always been many professional scientists with strong interests in the humanities, and vice versa, and it may also be that the strong sense of hostility between the two cultures is found more in Anglo-Saxon cultures than in others. Nevertheless, Snow had touched on something deep, of which the science wars represent only one, particularly ugly expression.

Somewhere amidst the sound and fury, we believe, is a deep, long-lived, and interesting set of problems. In this book we have brought together scholars to debate the issues in the hope of accomplishing a number of things. First, we want to reveal, through the style of the arguments themselves, that there is something going on here that is worthy of serious discussion, and furthermore—to answer the question posed by the book's title "The One Culture?"—that it *is* possible for practitioners and observers of science to meet on common ground. Second, we want to move the debate ahead in ways we will discuss below. We are quite sure we have accomplished the first aim; whether we have accomplished the second, the reader must decide.

For those who are new to these controversies, some background may be helpful. Our main concern is not the history of the debate, but its contemporary transformation into the conflict between natural scientists and those nonscientists who take a professional interest not so much in the findings of scientific knowledge, but in the workings and nature of science. Such studies have traditionally been divided into three fields—history of science, philosophy of science, sociology of science—although more recently there have been some attempts at unifying them under the more general heading of "science studies."

Prior to the 1960s there was little to bring these fields into conflict with the ranks of professional scientists. A substantial fraction of the history of science was written by scientists themselves, retired or moonlighting, and an even more substantial fraction was of a celebratory nature. Philosophy of science of course has a very long tradition—indeed is not easily separable from philosophy in general—but much philosophy of science was an attempt to explain the success of science rather than to challenge the scientific worldview, and so scientists were unlikely to find it disagreeable. Sometimes, of course, philosophical analysis has made the world of science less comfortable for science, but whatever the flavor, philosophy was largely ignored by scientists. As the aphorism cited by Steven Weinberg puts it: "philosophy of science is just about as useful to scientists as ornithology is to birds."[1]

Pre-1970s sociology of science examined the social institutions of science—the norms of acceptable scientific behavior, what motivates scientists, how they avoid bias, and the like. While by no means everything in this so-called Mertonian tradition (after Robert Merton, one of the main founders of the field) was entirely flattering to scientists' self-images, by and large the focus was on how these institutions enable science to work as well as it does. Again, no threat was perceived and members of the Merton school were welcomed into the institutions of science.[2]

In the 1960s, however, studies of science began turning in new directions. A major event was the appearance of Thomas Kuhn's *The Structure of Scientific Revolutions* (1996). Though first published in 1962, its full effects were not to be felt for a few years, and the details of Kuhn's influence on subsequent work have been much debated. What is certain is that he emboldened many investigators to expand their horizons by showing that natural science could itself be thought of as a kind of culture-building exercise.[3]

Subsequently, from the very early 1970s on, sociologists began turning their attention to the content, not just the institutions, of science. The new discipline of "sociology of scientific knowledge" was born, and systematic programs were established, such as the Edinburgh Strong Programme and the Bath school. The work of these groups stressed the cultural basis of scientific knowledge and the idea that similar experiments and theories could be differently interpreted so as to yield different scientific conclusions. Likewise, history of science became more professionalized, and less celebratory. And apparently foreign fields—literary criticism, cultural theory, feminist studies, etc.—began to incorporate scientific language and concepts into their work, or even made scientific texts and practices their prime object. These new directions in the study

1. It may have been Richard Feynman who first said this.

2. For example, onto the editorial board of the journal *Science*.

3. In the preface to his book, Kuhn cites a work that influenced him: Ludwik Fleck's *Genesis and Development of a Scientific Fact* (1979), first published in German in the 1930s. Fleck's book did not become widely known in the English-speaking world until the 1970s, but in terms of content, if not of direct influence, it is the true precursor of the sociology of scientific knowledge.

of science were perceived by many as challenging traditional conceptions of the scientific enterprise and elicited critical responses—some very vigorous—but primarily from other members of the science studies communities, notably philosophers.

WHAT IS AT ISSUE?

To characterize concisely the challenges posed by the new science studies is not easy. To simplify, we might represent them in terms of opposed conceptual clusters—for example, realism/rationalism/objectivism versus relativism/constructivism/subjectivism. The "new" science studies (now a quarter-century old) generally emphasize the latter cluster. They focus on the role of human factors in science and how scientific knowledge is *contingent* on and *constructed* by the operation of these factors—the social character of scientific institutions, the culture in which scientific investigation takes place, the language used to express scientific findings, etc. The contrasting position—held by the vast majority of scientists themselves as well as more traditional science studiers—places greater stress on how scientific knowledge is *determined* by the natural world and *codified* by the objective examination of that world.

Participants on both sides of the debate, however, occupy a very wide range of philosophical positions and commitments, often differing on points of great subtlety. This is worth remembering because much of the heat in the debate has been generated by failing to see the subtleties. We have been lucky enough to participate in a few discussions where what had seemed irresolvable conflicts turned out to depend on a misinterpretation of a word or a phrase, or a misunderstanding of the context. This occurred, for instance, in 1997 at a Science Peace Workshop in Southampton, where a group of science studiers and scientists assembled for three days of discussion, culminating in a public forum. Most participants agreed that it had been a success, even if they didn't agree on all the substantive issues.

A question that often divides people is whether these new movements are antithetical to science and reason. There are two dimensions to this issue: intent and consequences. Are the new critics consciously hostile to science? It has been claimed that they seek to reduce science's authority and influence in order to further a political interest, or even simply out of jealousy over the success and stature that scientists have achieved. The contrary view is that the new critics are not "criticizing," they are developing a "critique"—a neutral analysis of why and how science occupies its prominent place in today's world—at worst, misguided or irrelevant but harmless; at best, stimulating new ways of thinking about difficult problems.

If the latter is the correct way of looking at the matter, the new studies no more criticize science than the theater critic "criticizes" the theater or the music critic "criticizes" music. It is commentary and analysis that is the object, not an attack on the whole art form. Of course, the analogy should also remind us that artists and critics are not always the best of friends; those who cannot "do" but still take it upon themselves to analyze should tread carefully. In the sciences resentment is especially easy to understand as traditionally there has been no equivalent to the theater or the music critic; the sciences have been taken to be too difficult or esoteric to allow outside observers to have anything sensible to say to those who inhabit the "Republic of Science." It has come as somewhat of a shock to discover that outsiders can accomplish something like a critique; it seems obvious that the authority of the scientific community must be a little diminished by admitting outsiders and taking them seriously.

This takes us on to the question of consequences. Many of the opponents of this kind of science studies note with alarm contemporary trends such as decreasing funding for science, increasing suspicion of scientists, weakening of intellectual standards, widespread belief in creation science or astrology, and the like. They claim these trends are to a significant degree encouraged by constructivist/postmodern thinking in contemporary academia. The new critics respond by questioning whether science is truly at risk, noting the vast preponderance of positive works about science that can be found on the shelves of any bookshop, the small sales of books on science studies as compared to the much larger circulation of the works of their antagonists, and the relative power of the two sides in our universities and in public life. They also see no evidence of a causal connection between the new approaches and the above-mentioned trends—trends which they too decry, and which endanger science studies as much as they endanger science. They find at least as much reason to believe that disenchantment with science has grown out of attacks from the religious right and from some disappointment with what science has achieved compared to what it has promised since the Second World War. Indeed, some of the new critics believe it is their job to *defend* science by generating a more realistic model of what it can deliver in the difficult circumstances of disputed technological decision-making.

WHY THIS BOOK AT THIS TIME?

Unfortunately the science wars rumble on, and while more books have appeared (Gross, Levitt, and Lewis 1996; Ross 1996a; Koertge 1998), they have been one sided, consisting largely of statements of position with little or no reference to the counterarguments. We thought we should try to arrange something closer to an exchange of views. As a result of talking things through, as at the Southampton workshop mentioned above, we have, occasionally, seen academic antagonists start to see the point of each other's argument.

We hope we have advanced the debate in several ways. While we never expected that the outcome of this exercise would be complete agreement, we did want to bring about a little convergence—and we think we have. But debate can be made better even in the absence of convergence. We hope our efforts here will improve the way in which the disagreement is conducted in the future. We might define the difference between the pure forms of academic "war" and "debate" as that in the former the sole object is to win, whereas in the latter the aim is to convince your opponent. Winning, or appearing to win, without convincing your opponent can be done in a number of ways: you can talk to yourself; you can write for an audience that you think of as being neutral, which usually means easily convinced of your ideas; or you can aim at the mass media. All this can be accomplished by the use of arguments that would not convince even an open-minded opponent.

Second, without becoming obscure or boring we have tried to reintroduce some of the vital complexity that has become hard to see amidst the flash and smoke of bombs and shells. We would like to have brought about about a permanent change in this regard; we would like to make some of the more facile commentators think twice before they fire off a salvo; we want the "pundits" to consider that they may be revealing more of themselves than of their opponents if they are too quick to adopt a dismissive and haughty attitude.

The third thing we hope to have accomplished is to clarify, narrow down, and define the source of some of the unresolved differences. Unlike war, academic disagreement is not always bad. For the natural sciences as well as the social sciences and the humanities it serves as a motor of change; for the latter, it is even the stuff of undergraduate courses. Some of these disagreements may be hard or impossible to describe clearly—they may be a matter of looking at things from the point of view of different academic traditions so deeply ingrained as to be invisible. But there are other arguments that can be refined to the point that all parties know exactly what they disagree about and what they need to think about in order to take sides or move on.

It should be noted that we have chosen a rather narrow focus: we concentrate primarily on issues that have arisen out of the field called "sociology of scientific knowledge" (SSK) and the critical responses thereto. We invited scholars who have been active in SSK and closely related fields and scientists who have commented upon or (in one case) been involved in their projects to contribute. The decision to limit the scope of the book to SSK and its close neighbors was primarily a pragmatic one: we felt that to try to represent the whole of science studies and the full range of critics would lead at best to shallowness rather than the more careful engagement we were looking for. To gain even a little convergence in an argument of this sort is a difficult thing. The clearer the position the better, and we felt that if the social science authors were themselves in broad agreement we would have more chance of moving in the right direction.

This limitation should *not* be taken as tacit criticism of other movements in science studies, such as literary theory, actor-network theory, or cultural studies (although it is true that one of the editors is himself less happy about some of the other approaches than about SSK and has criticized some of them in print). One of the things that we hope the book will accomplish—to show that taking a comment or two out of context as the basis for an attack sets a dangerous precedent—should be seen as applying quite generally to all approaches. To ignore what seems implausible is not an unreasonable thing to do—life is short—but to attack it without any serious attempt to make contact with its proponents and understand the position being argued is academically inexcusable. (Michael Lynch discusses these concerns at greater length in chapter 29.)

On the science side, the book represents an equally narrow group: all but one of the active scientists are physicists. This may be a cause for some concern. The fact that physics has been by far the most common subject for philosophy and other science studies has raised questions over what can be said about universal "science" as opposed to a particular field of science. (Two of our contributors suggest that there might even be a systematic difference between the ways *theoretical* and *experimental* physicists think about these matters!) But this was not a deliberate decision; it is rather due to the fact that up to now the majority of scientists who have been paying serious attention to science studies have been physicists. (We did approach one or two nonphysicists, but they were unable to participate.) We hope things will evolve, and future stages in what we might call the Science Peace Process will include a more eclectic group of scientists; but even the physics specialists here exhibit significantly divergent views on a number of topics.

STRUCTURE, CONTENT, AND THEMES

The particular format we have chosen for the book tries to reflect the successful give-and-take structure of the Southampton workshop. Initially twelve authors (some with coauthors), drawn equally from the ranks of science studiers and practicing scientists, contributed position papers on a theme of their choice appropriate to the overall topic. The papers were then distributed, and everyone was invited to comment on whatever aspects of the papers evoked their interest or criticism. In the third and final round, authors were given the opportunity to respond to these commentaries and to defend, clarify, or even modify their positions.

This book does not unroll linearly. As a number of contributors point out, there is no easy division of positions into "sides"—one even talks of "strange bedfellows"—and the commentaries and rebuttals (also not unexpectedly) go in all directions. Some of them focus directly on just one or two major points of contention; others address a number of smaller issues; while still others introduce new material, inspired by rather than responding to earlier pieces.

How then should the material be organized, and how should it be read? It would be very difficult, and we will not even try, to provide a detailed road map to the wide-ranging set of topics and arguments. But we *will* try to offer the reader a little guidance by several means. We start by dividing the pieces into three sections corresponding to the three rounds of contributions; most arguments will be easier to follow if the three sections are read sequentially. Though the order in which pieces are read within each section is less important, we have further divided the first group, the original position papers, into four subgroups that seem to us to have some thematic coherence. It should be emphasized, though, that all of these first-round papers range across *all* the themes to at least some extent.

The first subgroup, "Philosophies," tend to focus on the issues at a fundamental level. Trevor Pinch's discussion of the intellectual antecedents of SSK as well as the science wars may serve as a useful starting point for many readers. Jean Bricmont and Alan Sokal then challenge the philosophical and methodological underpinnings of the SSK approach, while Michael Lynch's essay offers a defense.

Subgroup 2, "Perspectives," represent points of view that are somewhat off the main axis of the debate. Jane Gregory and Steve Miller, unlike most of the protagonists, approach the problem from an outsider's position, looking from the point of view of the public who have professional expertise in neither science nor science studies. Peter Saulson, in contrast, occupies a distinctively *inside* position: a scientist who has been the subject of an SSK case study.

In the third subgroup we return to the origins of the debate, but here the emphasis is somewhat more rhetorical than philosophical. David Mermin illustrates, out of his own experience, how easily misreadings and misunderstandings inflate into vehement disagreement, while Steven Shapin defends SSK against the charge of being "antiscientific."

The last subgroup are related by the substantial attention they give to future directions that might be followed in working out these issues. Steven Weinberg explores the use of historical understanding for the practice of science, and the converse. Peter Dear introduces a new term, "epistemography," as a descriptor to help understand the science studies program and its contributions. Kenneth Wilson and Constance Barsky propose a hitherto relatively neglected topic, the connection between scientific progress and improvement of accuracy, as potential grist for SSK's mill. Harry Collins argues for the inherent value of examining one's most basic, commonsense beliefs from a decidedly foreign perspective. And Jay Labinger sounds a similar theme in suggesting that practicing scientists may well benefit from participating in, not merely tolerating, science studies.

While most of the pieces in rounds 2 and 3 of the debate address issues and topics raised in earlier rounds, each can stand very well on its own as an independent essay and can be read that way, in any sequence (we have followed traditional, if unimaginative, alphabetical ordering). For the benefit of readers seeking to trace the evolution of particular themes through the debate—backwards, forwards, or sideways—we have provided a cross-referencing scheme, consisting of bracketed notations guiding the reader to the appropriate chapter. Thus an insertion of [15] following a specific point indicates that a response to, or an antecedent of, that point can be found in chapter 15.

Following all the individual essays, in the conclusion, we try to sum up by identifying some of the main themes of consensus and disagreement and offer our perspective on their significance in the wider picture.

Part One

POSITIONS

Chapter 2

DOES SCIENCE STUDIES UNDERMINE SCIENCE? WITTGENSTEIN, TURING, AND POLANYI AS PRECURSORS FOR SCIENCE STUDIES AND THE SCIENCE WARS

Trevor Pinch

THE TWO CULTURES IN CAMBRIDGE, 1939

In 1939 Ludwig Wittgenstein, one of the greatest philosophers of the twentieth century, possibly ever, at the age of forty-nine finally got his Chair at Cambridge and proceeded to give a series of lectures titled "Foundations of Mathematics." Remarkably, at Cambridge at the same time was one of the greatest mathematicians of the twentieth century, possibly ever, Alan Turing, who happened to be giving a series of identically titled lectures. Even more remarkably, Turing, who at twenty-seven was rather junior to Wittgenstein, decided to attend Wittgenstein's class.

This essay was adapted from a lecture delivered at Duke University in March 1997 as part of the colloquium series "Reconfiguring the Two Cultures." A version of the lecture was also delivered at the Science Peace Workshop held at Southampton University in July 1997 and at a conference on "Communicating Science," Hamilton College, October 1998. I am grateful for comments from Ray Monk, Arkady Plodnitsky, Harry Collins, and Jay Labinger.

We are lucky that the encounter between Wittgenstein and Turing is documented in two excellent intellectual biographies—I refer to Alan Hodges's *Alan Turing: The Enigma of Intelligence* (1983) and Ray Monk's *Ludwig Wittgenstein: The Duty of Genius* (1990). I will draw upon both books to tell the story of this encounter.

Hodges sets the scene:

> Sharing a brusque, outdoor, spartan, tie-less appearance (although Alan remained faithful to his sportsjacket, in contrast to the leather jacket worn by the philosopher), they were rather alike in this intensity and seriousness. Neither could be defined by official positions . . . for they were unique individuals, creating their own mental worlds. They were both interested only in fundamental questions, although they went in different directions. But Wittgenstein was much the more dramatic figure. Born into the Austrian equivalent of the Carnegies, he had given away a family fortune, spent years in a village school-teaching, and lived alone for a year in a Norwegian hut. And even if Alan was son of Empire, the Turing household had precious little in common with the Palais Wittgenstein. (153)

There is little doubt from Hodges's account that Turing cut a rather odd figure, but Wittgenstein was even odder.[1] Turing, before being accepted into the class, had to meet privately with Wittgenstein in his austere Trinity room. Turing described Wittgenstein as "'a *very* peculiar man,' for after they had talked about some logic, Wittgenstein had said that he would have to go into a nearby room to think over what had been said" (Hodges 1983, 153). Anyway, Turing made the cut and was accepted into Wittgenstein's class.

Hodges continues: "Alan joined Wittgenstein's class on Foundations of Mathematics. Although this had the same title as Alan's course it was altogether different. The Turing course was one on the chess game of mathematical logic; extracting the neatest and the tightest set of axioms from which to begin, making them flower according to the exact system of rules into the structure of mathematics, and discovering the technical limitations of that procedure (152)."

What was Wittgenstein's class about? We turn to Monk for elucidation:

1. Of course, Turing and Wittgenstein were both gay as well, but what bearing this had on the debate is not clear.

> The lectures on mathematics form part of Wittgenstein's general attack on the idol-worship of science. Indeed, he perceived this particular campaign as the most important part of that struggle. . . .
>
> . . . Wittgenstein's technique was not to reinterpret certain mathematical proofs, but, rather, to redescribe the whole of mathematics in such a way that mathematical logic would appear to be the philosophical aberration he believed it to be, and in a way that dissolved entirely the picture of mathematics as a science which discovers facts about mathematical objects (numbers, sets, etc.). "I shall try again and again," he said, "to show that what is called a mathematical discovery had much better be called a mathematical invention." There was, on his view, nothing for the mathematician to discover. A proof in mathematics does not establish the truth of a conclusion: it fixes rather the *meaning* of certain signs. The "inexorability" of mathematics, therefore, does not consist in *certain knowledge* of mathematical truths, but in the fact that mathematical propositions are *grammatical.* (416, 418)

Wittgenstein was opposed to the search for the foundations of mathematics, famously remarking that "foundations are no more the foundations for us than the painted rock is the support of the painted tower" (Monk 1990, 417–18). His lectures were also directed against those who searched for "hidden contradictions" in mathematics. Wittgenstein could not see how such contradictions could ever trouble a working mathematician and thus the search for them was pointless. His views derived from his wider schema for treating activities like mathematics as "language games." Understood in this way, it means that extraphilosophical justifications for the activity are unnecessary and potentially misleading.

Monk continues:

> Wittgenstein presumably thought that if he could persuade Turing to see mathematics in this light, he could persuade anybody. But Turing was not to be persuaded. For him, as for Russell and for most professional mathematicians, the beauty of mathematics, its very "charm," lay precisely in its power to provide, in an otherwise uncertain world, unassailable truths. . . . Asked at one point whether he understood what Wittgenstein was saying, Turing replied: "I understand, but I don't agree that it is simply a question of giving new meanings to words." To this Wittgenstein—somewhat bizarrely—commented: "Turing doesn't object to anything I say. He agrees with every word. He objects to the idea he thinks under-

> lies it. He thinks we're undermining mathematics, introducing Bolshevism into mathematics. But not at all." (418–19)

After what he perceived to be several useless discussions, Turing dropped the class.[2]

It is important to note that this encounter took place in a secluded set of Cambridge college rooms. It did not take place in the *New York Times* op-ed pages, or in the *New York Review of Books,* or in a journal called *Social Text.* There was no Alan Sokal hoax fanning the flames, there were no interdisciplinary departments of science studies and cultural studies to be envious of, there was no crisis in the funding of physics, there was no multiculturalism or political correctness. There were just two very smart people—one in a sports jacket and one in a leather jacket—and they could not understand each other.

The moral I draw from this story is that the misunderstanding that arises between practicing scientists and those who adhere to some version of philosophical skepticism about science, as Wittgenstein did, is very deep, serious, and profound. In the midst of the current furor known as the "science wars," the issues often get trivialized to the detriment of both sides.[3]

FROM *THE TWO CULTURES* TO THE SCIENCE WARS

"The science wars" refers to a debate raging within the academy and without over the status of fields like science and technology studies and cultural studies of science and technology. The debate has largely been initiated by natural scientists who have written books and made public

2. Ray Monk has pointed out (in private communication) that Wittgenstein was not only trying to redress the philosophic pretensions of mathematics but actually trying to change parts of pure mathematics. In this regard, it appears that Wittgenstein was a more radical philosophical skeptic that most people working in the sociology of scientific knowledge would want to be. What Wittgenstein was really up to is contested territory. The reading of Wittgenstein on science and mathematics to which I subscribe is that taken from David Bloor, who argues that mathematics and science are both forms of life and that key activities like learning to follow rules and solve problems are to be treated as sociological matters. See, for instance, Bloor 1973 and 1983. For another extension of Wittgensteinian ideas to the sociological study of science, see Collins and Pinch 1982.

3. Part of the trivialization comes from labeling complex intellectual positions as "the science studies approach" or "postmodern science." It is worth noting that there are many ways to be a philosophical skeptic about science and there are, as noted above, different degrees of skepticism; there are also many routes to these positions and many differences and nuances among people holding such positions.

statements critical of science studies and what they take to be some of its central ideas.[4]

A book of which I am a coauthor, *The Golem: What You Should Know about Science* (Collins and Pinch 1993, 1998a; see also Collins and Pinch 1998b), which is an attempt to popularize the findings of the sociology of science, has itself become part of the wider debate. A well-known physicist, David Mermin, devoted two columns (1996a, 1996b) in the American physicists' house journal, *Physics Today,* to an examination of the Golem's argument and has thereby generated a large amount of correspondence. Mermin's treatment is critical but written in a recognizably academic register. Other physicists have been less temperate. One described the book as "silly" and suggested that the only reason the authors could have had for writing such a book is that they were "on the make" or "to get tenure" (Evans 1996). In short, misunderstandings and distrust are widespread. The space between the two cultures seems less a gap and more a yawning chasm.

The idea of the two cultures was made famous by C. P. Snow in the 1950s. It has come to refer to the separate self-contained cultures of the humanities and of the natural sciences. The two cultures were held to be largely ignorant of each other. What is perhaps surprising is why the two cultures have been able to coexist peacefully for so long. Aside from one or two skirmishes, such as that between Wittgenstein and Turing, the science wars are more the exception than the rule.

It is important to bear in mind that radical disjunctions in our cultural sensibilities and practices need not necessarily lead to clashes. As long as each culture can flourish, receive ample material resources and symbolic legitimation, there need be little reason for dissent. In the case of science and the humanities, elites in both cultural domains have, on the whole, tended to sing the praises of each other. Statements of approbation flow both ways. Where there is overlap we find admiration expressed for the individual creative geniuses on both sides. Where there is difference it is not so much a source of dispute, but rather a matter of each agreeing to pursue different sorts of activity, whether it be scientists exercising their genius in understanding nature or being useful in solving real world problems or humanists tapping into their wells of creativity in literature and the arts or providing food for the soul. A famous scientist, like Richard Feynman, can do his science as well as take up the

4. Several natural scientists have published attacks on science studies or aspects of it. The most notorious of these is Gross and Levitt 1994. For a reply to some of Gross and Levitt's most egregious errors, see Hart 1996. For a reading of the science wars debate more sympathetic to science studies, see Labinger 1997.

bongos and the paint brush and go on to endorse art. Art is indeed a worthy human activity, but as Feynman reminds us, it is a different activity and should not be confused with the business of doing science (1986).

I suggest that one reason for the emergence of the science wars is that the field of science studies breaks down this cozy relationship. Rather than treating science as the "exotic other" or just as a different animal, it levels the playing field—all animals are really the same, and they are not all that exotic. Within science studies, science is treated as another body of skilled practice, not unlike other areas of human endeavor. Science studies, rather than endorsing the parallelism of the two-culture thesis, asserts that there is *one culture* in the sense that both science and the humanities share the characteristics of being cultures of expertise.

THE SOCIOLOGY OF SCIENTIFIC KNOWLEDGE

It was not until 1973 that the full implications of Wittgenstein's remarks on the nature of mathematics were developed for the study of science (Bloor 1973, note 4). David Bloor's Strong Programme in the sociology of science showed how the Wittgensteinian approach to mathematics could be developed into a thoroughgoing and systematic sociology of scientific knowledge. The extension of the sociology of knowledge to the natural sciences was one of the crowning achievements of science studies in the 1970s and 1980s.

A fundamental concern of the sociology of knowledge is how different societies decide what counts as part of their taken-for-granted beliefs, "stocks of knowledge," and worldviews, and what counts as mistakes or erroneous beliefs. This is part of a much wider concern to understand how knowledge and taken-for-granted beliefs get socially embedded or institutionalized within a particular society. This endeavor seems straightforward when dealing with beliefs such as religion or when dealing with clearly failed bodies of knowledge such as alchemy. Bloor, however, argues that a fully symmetrical sociology of scientific knowledge must seek similar explanations for the institutionalization of successful areas of knowledge such as modern physics. The pursuit of such explanations, however, means that sociologists ask questions of science which to practicing scientists often seem plain daft, irreverent, or, worse, undermining of the scientific enterprise.

We ask not why science is more true than other accounts of the world or why the scientific method is the only valid method for generating truth, but rather how do scientists reach agreement about what *counts* as truth and as valid scientific method and how have these notions arisen historically? Asking these sorts of questions inevitably means putting aside epistemologically loaded terms like "truth." It inevitably means embracing some form of skepticism.

For most people working within the sociology of scientific knowledge this stance is a *methodological* precept (stemming from what David Bloor calls "symmetry").[5] If we assume in advance of analyzing any particular piece of science that we, as analysts, have independent access to, say, truth, then this is to beg the question of what is at stake and, worse, can lead to a form of social epidemiology whereby we only use social explanations for deviations from truth. We try to avoid explaining the emergence of truth by reference to its truthfulness because such explanations are circular. This methodological precept becomes almost unavoidable in studying, as many of us do, scientific episodes as they unfold, where the "truth" of matters has not even been established by the scientists. We think we find out more interesting things in the sociology and history of science with such a methodological precept.

WHAT'S AT STAKE IN UNDERMINING?

But establishing the sociology of scientific knowledge as an academic endeavor which asks legitimate questions about a major institution may not be enough. For instance, you might say that it is legitimate for criminologists to study crime patterns, the criminal "underworld," and how people get labeled as criminals, but this does not mean that you would feel comfortable if, having just been robbed, you ran into a criminologist who insisted upon explaining to you the social processes whereby people become labeled as criminals. There is no doubt what Wittgenstein perceived to be Turing's objection, that he was in some way *undermining* mathematics, is, for many, the nub of the issue. For many practicing scientists, such as Turing, the hallmark of why scientific knowledge is special and not to be treated on a par with other bodies of knowledge or other sorts of activities is precisely because science is epistemologically privileged. By bracketing truth and, worse, by comparing science with baseball, as Stanley Fish has recently suggested, or plumbing, as we he-

5. The methodological nature of symmetry is discussed in Pinch 1984.

retically suggest in *The Golem,* we seem to be giving the impression of undermining science.

It is this undermining aspect which is essential to grasp because it gets to the core of much of the misunderstanding leading to the science wars. Are people in science studies undermining science? I cannot speak for science studies as a whole, but certainly within the tradition of work associated with the Strong Programme and in *The Golem* there is *no intention* to undermine science. On the contrary, in *The Golem* our aim is explicitly to provide an account of science which we believe, in the long run, will help the public understand the nature of science and thereby support science. But claims, of course, can undermine despite the best of intentions. Thus our criminologist, by giving an exposition of labeling theory, may have simply wished to put your loss into a wider perspective, but he may still have served to undermine your status as a victim of crime.

Another instructive example of a body of expertise which has the potential to undermine is that of insurance actuaries. They make it their business to assess risk by assigning numerical values. From the point of view of the actuary, the worth of your life is reduced to the outcome of a set of calculations of varying degrees of risk. Most of the time we do not object to this way of going about things. We do not reject their expertise because it undermines the value of human life or claim that actuaries are charlatans, antilife, and so on. But in some circumstances the claims of actuaries could be taken to be undermining. Imagine a loved one has just died, and you are faced by an actuary who is only concerned with the financial rewards that your loved one's insurance policy will bring. In *this* context their way of valuing life could be taken to undermine your way of valuing life.

In assessing whether any set of claims can be said to undermine an endeavor, we are at the same time dealing with *who* is making these claims, towards *whom* they are directed, and *under what circumstances*. If a *physicist* says casually to a fellow physicist whom he meets in the elevator, "These experiments are messy, near the noise, and I don't know whom to believe," the chances are high that his colleagues would not regard him as undermining science. However, if a *sociologist of science* says in print, "These experiments are messy, near the noise, and sometimes the scientists don't know whom to believe," *that* is often seen as undermining science. Scientists tend to hear heresy when such claims are made by us in public rather than by them in private.

When it comes to assessing claims made by sociologists of science, the relevant audience may not be scientists at all. This is particularly true for work within the sociology of science which addresses the issue of the public understanding of science and which may be directed at the public itself. This is the case with the *Golem* series, where our aim has been to write for members of the public rather than for scholarly audiences. Paradoxically it is *The Golem*'s appeal to the wider public which most irks some physicists. Thus, one physicist has told me in no uncertain terms that if *The Golem* had been a straightforward scholarly account rather than a popularization, he would not have worried about it. Thus, in this case, the undermining comes not so much from *who* is making the claims but *to whom* such claims are directed and *how* such claims are presented.[6]

FLIP-FLOP THINKING ABOUT SCIENCE

Today there is a crisis over the public's understanding of science. This is reflected in recent political crises over matters to do with science and technology. There seems to be enormous instability in the relationship between science and its wider public. Part of this instability comes from the growing awareness on the part of the public that they can have a say in important decisions concerning science and technology, such as whether or not nuclear power is safe, or in Britain, with mad cow disease, whether or not to eat hamburgers.[7] The recent political crisis in Britain over the safety of genetically modified foods is the latest in a long line of crises where the public has received conflicting advice from experts as to the best course of action.

Conflicting advice is itself not the source of the instability. It is after all normal in life to face conflicting advice. The real instability comes from the public's unfamiliarity with science as being like everything else in life—an area where experts might disagree. The problem is that scientists have surrounded their enterprise with an aura of certainty. While scientists might be prepared to admit *in private* or to each other that science is human, involves skill, and contains uncertainties and that the idea of simple decisive experiments shooting down rogue theories is at best mythical or at worst a useful fiction for teaching purposes, *in public* they, by and large, maintain the view that science is about certainty.

The human side of science and its uncertain face are, however, con-

6. But here again the issue becomes more complicated. Intended readership might not match actual readership. It seems the main sales of *The Golem* have been to undergraduates for use in introductory science studies courses, not to the wider public.

7. Or as one wit put it: "To Beef or Not to Beef—That Is the Question."

stantly leaking through into public attention, whether through cases of fraud in science, overtrumpeted discoveries like cold fusion, or the sight of experts in disarray. The public increasingly sees how hard it is sometimes for scientists to reach agreement. Brought up on overweening promises of certainty, the public do not know how to handle uncertainty and disagreement. And they often overreact by totally rejecting science or by seeing it as a political conspiracy [17]. In *this* particular context, I suggest the science studies view, that science is a body of expertise with a human face to it, will not undermine the enterprise but actually will lead the public to respect science for what it is.

The aura of certainty with which science cloaks itself contrasts dramatically with the uncertainties to which the public are sometimes exposed. This leads to (as we put it in *The Golem*) "flip-flop thinking about science." For many people science is about certainty and scientists are seen as gods bringing down the sacred truths from the mountains of Nature. The danger of this view is that it can lead to disillusionment and the switch to an alternative view that sees science as a political conspiracy with scientific results fraudulently produced to satisfy cynical political masters.[8] What is needed are ways to understand the uncertain face of science which avoid the traps of falling into these two extreme views. The expertise model of science offered in science studies is one way out of the binary opposition of flip-flop thinking.

THE EXPERTISE MODEL

What then goes into expertise? One of the most important aspects of science to which the sociology of scientific knowledge has drawn attention is the craft or tacit knowledge involved in doing science. It was the British chemist and philosopher Michael Polanyi who first coined the term "tacit knowledge" to describe the nature of many experimental skills in science (1967). Tacit knowledge is knowledge which can be learned, transmitted, and passed on without being articulated. Tacit knowledge is passed on in "learning by doing."

Polanyi's favorite example was that of learning to ride a bike. In learning to ride a bike, explicit knowledge such as the physics of balance does not count for a lot. In practice when we teach someone to ride a bike,

8. A compelling example of this process is what happened in the Cumbrian sheep-farming communities after the farmers discovered that scientists claims about the affects of Chernobyl fallout being short lived were found to be wrong. The farmers turned to a cynical political conspiracy view of science. For discussion of this case, see Wynne 1996.

we do not go about it by sitting them down with a physics book. It is a practical task to be learned in the course of doing bike riding.

The skills acquired in many crafts like pottery and carpentry are learned on the job. Book knowledge will only take you so far, and Polanyi argued that the same was true of science, where of course a long apprenticeship at the lab bench is crucial. Cooking is another activity full of tacit skills, and again cooking is best learned not from a book, but from an apprenticeship. Apply for a job as a chef in a leading restaurant, and they will not want to know how many books on cooking you have read, but in whose kitchen you learned to cook. Cooking, say, a cheese soufflé is something which is notoriously hard to do unless you have learned a variety of skills, such as "folding in."

Sociologists have carried out studies of how skills are transmitted in science. For instance, Collins carried out important early research which followed how groups of scientists learned to build a new sort of laser first built by a Canadian group in the early 1970s (Collins 1974). Collins found that of all the other groups trying to build lasers, the only ones who could get their lasers to work had visited labs that already had a working laser. Papers distributed describing this new laser were alone not enough to enable scientists to get their lasers working. Collins interpreted his results as being another example of the role of tacit knowledge in science.[9]

There need be nothing particularly radical for scientists about this finding. Scientists themselves will tell you about the "golden hands"—the only person in the lab who can get a certain experiment to work. Graduate students know that the skills to get experiments to work often reside with the lab technicians.

Skill and who has it is important in understanding the process of science. It was Thomas Kuhn who said that a characteristic of science carried out within one paradigm or "normal science" is that if a scientist cannot solve a puzzle, it is the scientist who is usually taken to be at fault rather than the paradigm (1996). The same is true when an experiment has a known outcome. If scientists fail to get their laser to lase, the assumption is that they have not yet acquired the skills or competence to build a laser rather than that they have made a new discovery about laser physics. They just have to go at it a bit harder and acquire the necessary skills. And in building lasers there is a hard and fast outcome of

9. Collins also set the notion of tacit knowledge within the neo-Wittgensteinian framework of "forms of life." The "forms of life" idea is developed further in Collins and Pinch 1982.

success—your laser will either lase and vaporize a block of concrete or it will not. The same is true of most school science, where a premium is placed upon getting the "right" answer.

Although in much of science there is agreement as to the correct outcome of experiments, there are moments of scientific controversy where the "correct outcome" is what is at stake. These are cases where, for instance, replicating experiments claim different things and where experimental results seem to conflict with theory. Such cases have received a lot of attention within the history and sociology of science.[10] A recent dramatic example of such a case is the cold fusion controversy.

Skill and expertise are again the issue at stake in these sorts of cases. Where there is an agreed outcome, there is an agreed way of measuring skill—you can either ride your bike or you cannot, you can either get your laser to lase or you cannot. In a scientific controversy with clashing experimental results, a decision has to be made as to who has the requisite skill. Again, cookery is useful to turn to for illustration. Imagine you want to try out a new recipe which you see a chef demonstrate on television. You jot down the recipe, assemble the ingredients, and go at it. Alas your exotic mash turns to mush. Perhaps you missed some crucial ingredient? You call the station and get it all down; you may even substitute better ingredients. Again you fail. Now you have to make a judgment of competence and skill. Is it that the new dish is no good and that the chef is incompetent in this case, or is it you who are incompetent? You might bow to the superior qualifications of the TV chef and conclude it was a good dish but you just weren't up to it. Or you could go the other way. You could decide, "Darn it, I've been cooking for twenty years and I really know my stuff—also those TV chefs, they always cheat a bit. Perhaps the dish was just no good!"

The sort of dilemma you face is similar to that facing scientists who have to make judgments about whom to believe and whom not to believe during a scientific controversy. It is rarely a straightforward matter deciding these things. All sorts of factors might become relevant: you might appeal to theory, for example, "This claim is just too implausible to believe"; you may appeal to prestige, for example, the joke in the cold fusion case made by the skeptics at Caltech and MIT was that only labs with football teams saw cold fusion. (This says something about the pecking order in US physics departments.) Often a mixture of factors will be used to decide who has the requisite skill and who does not.[11]

10. See, for instance, Collins 1975, Collins and Pinch 1982, Collins 1981c, Rudwick 1985, Shapin and Schaffer 1985, Pickering 1984, Pinch 1986, and Collins 1992.

11. For a discussion of the role of humor in closing the cold fusion controversy, see Pinch 1995.

The point here can be made in an even more fundamental way. When claims about the natural world are made, there are always issues of credibility and trust at stake as well. Nature doesn't speak for him or herself. There is always a scientist speaking on behalf of nature and we have to decide who is credible and trustworthy and who isn't. Mostly this process will be invisible—as invisible as your trust in your banker when you deposit your money in the bank, that is until the bank is threatened with collapse and trust once more appears as an issue. The same happens during a scientific controversy. It is not that you have nature on one side and social relations on the other. We could make no claims about nature at all without a network of trusting social relationships within which claims about nature are to be judged.[12]

Cases of scientific controversy are perhaps the most important for the public's understanding of science. It is in learning to deal with cases of divided expertise, where there is controversy, where the public most needs help. In almost all the cases where instability and political crisis between scientists and their publics have arisen, there is conflicting expert advice. Work in science studies cannot tell the public which expert is correct, but it can help the public realize that uncertainties and disagreement amongst technical experts is sometimes the name of the game.

The view of science as a body of expertise stresses that scientists are experts who reach their conclusions sometimes in a messy way. Scientific experts are skillful practitioners like other skilled groups in society, like potters, carpenters, real estate agents, and plumbers. We should respect such groups as experts. Just because a carpenter does not claim absolute certainty does not mean that she cannot make a beautiful cabinet that works perfectly. As we note in *The Golem,* because no one thinks of plumbing as being immaculately conceived, we do not have anti-plumbing movements in our society [17].

It is important to stress that the claim is not that science is the same as plumbing or cooking, but rather that the *right model for thinking about the nature of science in relation to its publics and in relation to other institutions* is to think of it as a body of expertise carried out by human practitioners. Science should be accorded all the attention and respect we give other experts in our society.

In comparing science with plumbing and cookery, have I undermined science? I hope not. There is surely no doubt that cookery works, and there is no doubt that science works. Treating cookery as a skilled practice shaped by culture with its own history does not stop anyone

12. This point is extensively argued and documented by Shapin (1994).

enjoying a fine meal. The same is true of science. Treating science as a social product does not, and should not, mean that we cannot appreciate, say, relativity theory or Millikan's achievements in measuring the charge of the electron as fine and useful human achievements.

It was the perception of "undermining" which according to Wittgenstein led to his and Turing's failure to understand each other. Wittgenstein's agenda for science and that of science studies both seem to engender the same reaction on the part of practicing scientists. If nothing else, the science wars have provided an opportunity to open a dialogue which should have started sixty years ago in Cambridge.[13] Some of us have tried recently with moderate success to turn war war into jaw jaw.[14] This volume hopefully extends this endeavor.

13. Another important attempt to bring the two sides together has been Arkady Plotnitsky's heroic efforts to demonstrate that physicists such as Alan Sokal's reading of certain remarks by Derrida are actually misreadings (Plotnitsky 1997).

14. For example, the Southampton Science Peace Workshop brought together scientists and academics in science studies for three days of civil debate. Also, the tone of some recent exchanges in the April 1999 issue *Social Studies of Science* concerning articles in A House Built on Sand (Koertge 1998) is rather encouraging.

Chapter 3

SCIENCE AND SOCIOLOGY OF SCIENCE: BEYOND WAR AND PEACE

Jean Bricmont and Alan Sokal

1. INTRODUCTION: NEITHER WAR NOR PEACE

When we were asked to contribute to this volume, our immediate reaction was: We did not want a "science war" in the first place, so why should we now want peace? We don't want a truce either. These are simply not the right categories for an intellectual discussion. In "peace talks," one can and must negotiate: I'll give you this and you give me that. But truth cannot be negotiated in this way. In fact, adopting this "diplomatic" terminology would amount to conceding too much to the relativist philosophy—for which intellectual discussion is nothing more than a power struggle involving a mixture of persuasion, coercion, and negotiation—that we want to criticize. On the other hand, we do believe in exchanges of ideas: these serve to clarify areas of agreement and disagreement, to test theses by subjecting them to objections, and more generally to promote a collective search for the truth. And that is why we are here.

For us it all started a few years ago: we were both puzzled and irritated by the prevailing philosophy in certain intellectual circles—encompassing large parts of the humanities, anthropology, and sociology of science—whereby all facts are "socially constructed," scientific theories are mere "myths" or "narrations," scientific debates are resolved by "rhetoric" and "enlisting allies," and truth is simply intersubjective agreement. If all this seems an overstatement, consider the following assertions:

> [T]he validity of theoretical propositions in the sciences is in no way affected by factual evidence. (Gergen 1988, 37)

> The natural world has a small or non-existent role in the construction of scientific knowledge.[1] (Collins 1981a, 3)

> Since the settlement of a controversy is the *cause* of Nature's representation, not the consequence, we can never use the outcome—Nature—to explain how and why a controversy has been settled.[2] (Latour 1987, 99, 258; emphasis in original)

> For the relativist [such as ourselves] there is no sense attached to the idea that some standards or beliefs are really rational as distinct from merely locally accepted as such. (Barnes and Bloor 1981, 27; clarification added by us)

> Science legitimates itself by linking its discoveries with power, a connection which *determines* (not merely influences) what counts as reliable knowledge. (Aronowitz 1988, 204; emphasis in the original)

Out of this irritation came Sokal's *Social Text* parody article and our book (Sokal 1996b; Sokal and Bricmont 1998).[3] Rather than repeat here all the arguments contained in our book and in other publications,[4] we shall try to summarize our main objections to the trends in science studies (or sociology of scientific knowledge) loosely inspired by the Strong Programme. These objections are epistemological and methodological; therefore, we will not discuss here the details of "case studies." While we do not deny that interesting work may have been done in those studies—particularly when the authors violated their own declared methodological precepts[5]—this fact does not answer our objections aimed at the principles of the Strong Programme [15].

1. Two qualifications need to be made: First, this statement is offered as part of Collins's introduction to a set of studies (edited by him) employing the relativist approach and constitutes his summary of that approach; he does not *explicitly* endorse this view, though an endorsement seems implied by the context. Second, while Collins appears to intend this assertion as an empirical claim about the history of science, it is possible that he intends it neither as an empirical claim nor as a normative principle of epistemology, but rather as a methodological injunction to sociologists of science: namely, to act *as if* "the natural world ha[d] a small or non-existent role in the construction of scientific knowledge," or in other words to *ignore* ("bracket") whatever role the natural world may in fact play in the construction of scientific knowledge. We shall argue in section 3.2 below that this approach is seriously deficient *as methodology* for sociologists of science.

2. See Sokal and Bricmont 1998, chapter 4, for a detailed discussion.

3. Though deliberate satire may seem to be a warlike tactic, we should stress that the main reason for using this tool was to force an honest debate which seemed blocked.

4. See Sokal 1998 and Bricmont 2001. For related arguments, see Nagel 1997, Haack 1998, and Kitcher 1998.

5. See Gingras 1995 for a critical analysis of several recent tendencies in science studies.

Over the last three years, we have participated in numerous debates with sociologists, anthropologists, psychologists, psychoanalysts, and philosophers. Although the reactions were extremely diverse, we have repeatedly met people who think that assertions of fact about the natural world can be true "in our culture" and yet be false in some other culture.[6] We have met people who systematically confuse facts and values, truths and beliefs, the world and our knowledge of it. Moreover, when challenged, they will consistently deny that such distinctions make sense. Some will claim that witches are as real as atoms or pretend to have no idea whether the Earth is flat, blood circulates, or the Crusades really took place. Note that these people are otherwise reasonable researchers or university professors. All this indicates the existence of a radically relativist academic Zeitgeist, which is weird.[7] To be sure, these are oral statements made in seminars or private discussion, and oral statements usually tend to be more radical than written ones. But the published assertions quoted in the preceding paragraph are already quite weird [23].[8]

If one inquires about the justifications for these surprising views, one is invariably led to the "usual suspects": the writings of Kuhn, Feyerabend, and Rorty; the underdetermination of theories by data; the theory-ladenness of observation; some writings of (the later) Wittgenstein; the Strong Programme in the sociology of science. Of course, the latter authors do not usually make the most radical claims that we have heard. Rather, what typically happens is that they make ambiguous or confused statements that are then interpreted by others in a radically relativist fashion. Therefore, our goal in this article will be to disentangle various confusions caused by fashionable ideas in the contemporary philosophy of science. Roughly speaking, our main thesis is that those ideas contain a kernel of truth that can be understood properly when those ideas are carefully formulated, but then they give no support to radical relativism.

However, before getting to work, we want to avoid possible misunderstandings and to emphasize some points which, we hope, are noncontroversial:

6. For an example involving the origins of Native American populations, see Sokal and Bricmont 1998, epilogue, and Boghossian 1998.

7. We emphasize that we have no idea how widespread these extreme positions are. But their mere existence is weird enough.

8. For extremely weird written statements, see also the discussion by Latour of the causes of the death of the pharaoh Ramses II (Latour 1998); and for a critique, see Sokal and Bricmont 1998, note 123.

1. Science is a human endeavor, and like any other human endeavor it merits being subjected to rigorous social analysis. Which research problems count as important; how research funds are distributed; who gets prestige and power; what role scientific expertise plays in public-policy debates; in what form scientific knowledge becomes embodied in technology, and for whose benefit—all these issues are strongly affected by political, economic, and to some extent ideological considerations, as well as by the internal logic of scientific inquiry. They are thus fruitful subjects for empirical study by historians, sociologists, political scientists, and economists.

2. At a more subtle level, even the content of scientific debate—what types of theories can be conceived and entertained, what criteria are to be used for deciding between competing theories—is constrained in part by the prevailing attitudes of mind, which in turn arise in part from deep-seated historical factors. It is the task of historians and sociologists of science to sort out, in each specific instance, the roles played by "external" and "internal" factors in determining the course of scientific development. Not surprisingly, scientists tend to stress the "internal" factors while sociologists tend to stress the "external," if only because each group tends to have a poor grasp on the other group's concepts. But these problems are perfectly amenable to rational debate.

3. There is nothing wrong with research informed by a political commitment, as long as that commitment does not blind the researcher to inconvenient facts. Thus, there is a long and honorable tradition of sociopolitical critique of science, including antiracist critiques of anthropological pseudoscience and eugenics and feminist critiques of psychology and parts of medicine and biology.[9] These critiques typically follow a standard pattern: First one shows, using conventional scientific arguments, why the research in question is flawed *according to the ordinary canons of good science;* then, *and only then,* one attempts to explain how the researchers' social prejudices (which may well have been unconscious) led them to violate these canons. Of course, each such critique has to stand or fall on its own merits; having good political intentions doesn't guarantee that one's analysis will constitute good science, good sociology, or good history. But this general two-step approach is, we think, sound; and empirical studies of this kind, if conducted with due

9. We limit ourselves here to critiques challenging the substantive content of scientific theories or methodology. Other important types of critiques challenge the uses to which scientific knowledge is put (e.g., in technology) or the social structure of the scientific community.

intellectual rigor, could shed useful light on the social conditions under which good science (defined normatively as the search for truths or at least approximate truths about the world) is fostered or hindered.[10]

Now, we don't want to claim that these three points exhaust the field of fruitful inquiry for historians and sociologists of science, but they certainly do lay out a big and important area. In any case, our criticisms of the sociology of scientific knowledge (SSK) are aimed only at radical epistemological and methodological claims that go far beyond the above-mentioned truisms.

This paper is organized as follows: First of all, we have to clear the epistemological ground (section 2). Next, we formulate our objections to the Strong Programme (section 3). And finally, we shall indicate what we think are some possible areas of collaboration between scientists and sociologists of science (section 4).

2. SOME BRIEF EPISTEMOLOGICAL REMARKS

2.1. Radical Skepticism, Underdetermination, and All That

Before discussing some serious issues in the philosophy of science, we need to clear out of the way some old red herrings. The first point, which should be noncontroversial, is that solipsism (the idea that there is nothing in the world except my sensations) and radical skepticism (that no reliable knowledge of the world can ever be obtained) cannot be refuted. It is doubtful whether anyone really believes those doctrines—at least when crossing a city street—but their irrefutability is nevertheless an important philosophical observation. Since the arguments are standard and go back at least to Hume, we need not repeat them here. Unfortunately, many of the arguments adduced in favor of relativist ideas are, in reality, banal reformulations of radical skepticism applied in unjustifiably selective ways.[11]

In the same way that nearly everyone in his or her everyday life disregards solipsism and radical skepticism and spontaneously adopts a "real-

10. Of course, we don't mean to imply that the *only* (or even principal) purpose of the history of science is to help working scientists. History of science obviously has intrinsic value as a contribution to the history of human society and human thought. But it seems to us that history of science, when done well, can *also* help working scientists.

11. Another favorite tactic employed by relativists is to conflate facts and our knowledge of them, not by giving any argument, but simply by using intentionally ambiguous terminology. See Sokal and Bricmont 1998, chapter 4, for examples in the works of Kuhn, Barnes-Bloor, Latour, and Fourez.

ist" or "objectivist" attitude toward the external world, scientists spontaneously do likewise in their professional work. Indeed, scientists rarely use the word "realist," because it is taken for granted: *of course* they want to discover (some aspects of) how the world really is! And of course they adhere to the so-called correspondence theory of truth[12] (again, a word that is barely used): if someone says that it is true that a given disease is caused by a given virus, she means that, in actual fact, the disease is caused by the virus. Philosophers often regard such views as naïve, but we would like to show that they are actually quite defensible, with, however, some important qualifications.

The main objections to the scientists' spontaneous attitude consist of various theses showing that theories are underdetermined by data.[13] In its most common formulation, the underdetermination thesis says that, for any finite (or even infinite) set of data, there are infinitely many mutually incompatible theories that are "compatible" with those data. This thesis, if not properly understood,[14] can easily lead to radical conclusions. The scientist who believes that a disease is caused by a virus presumably does so on the basis of some "evidence" or some "data." Saying that a disease is caused by a virus presumably counts as a "theory" (e.g., it involves, implicitly, many counterfactual statements). But, if one is able to convince the scientist that there are infinitely many distinct theories that are compatible with those "data," he or she may well wonder on what basis one can rationally choose between those theories.

In order to clarify the situation, it is important to understand how the underdetermination thesis is established; then, its meaning and its limitations become much clearer. Here are some examples of how underdetermination works; one may claim that:

- The past did not exist: the universe was created five minutes ago along with all the documents and all our memories referring to the (alleged) past in their present state. Alternatively, it could have been created one hundred or one thousand years ago.
- The stars do not exist: instead, there are spots on a distant sky that emit exactly the same signals as those we receive.
- All criminals ever put in jail were innocent. For each alleged criminal, explain away all testimony by a deliberate desire to harm the accused;

12. We would not even call it a "theory"; rather, we consider it a *precondition for the intelligibility* of assertions about the world.

13. Often called the Duhem-Quine thesis. In what follows, we will refer to Quine's version (1980), which is much more radical than Duhem's.

14. Particularly concerning the meaning of the word "compatible." See also Laudan 1990b for a more detailed discussion.

> declare that all evidence was fabricated by the police and that all confessions were obtained by force.[15]

Of course, all these "theses" may have to be elaborated, but the basic idea is clear: given any set of facts, just make up a story, no matter how ad hoc, to "account" for the facts without running into contradictions.[16]

It is important to realize that this is all there is to the general (Quinean) underdetermination thesis. Moreover, this thesis, although it played an important role in the refutation of the most extreme versions of logical positivism, is not very different from the observation that radical skepticism or even solipsism cannot be refuted: all our knowledge about the world is based on some sort of inference from the observed to the unobserved, and no such inference can be justified by deductive logic alone. However, it is clear that, in practice, nobody ever takes seriously such "theories" as those mentioned above, any more than they take seriously solipsism or radical skepticism. Let us call these "crazy theories."[17] (Of course, it is not easy to say exactly what it means for a theory to be noncrazy.) Note that these theories require no work: they can be formulated entirely a priori. On the other hand, the difficult problem, given some set of data, is to find even one noncrazy theory that accounts for them. Consider, for example, a police inquiry about some crime: it is easy enough to invent a story that "accounts for the facts" in an ad hoc fashion (sometimes lawyers do just that); what is hard is to discover who really committed the crime and to obtain evidence demonstrating it beyond a reasonable doubt. Reflecting on this elementary example clarifies the meaning of the underdetermination thesis. Despite the existence of innumerable "crazy theories" concerning any given crime, it sometimes happens in practice that there is a unique theory (i.e., a unique story about who committed the crime and how) that is *plausible* and compatible with the known facts; in that case, one will say that the criminal has been discovered (with a high degree of confidence, albeit not with certainty). It may also happen that no plausible theory is found, or that we are unable to decide which one among several suspects is really guilty: in these cases, the underdetermination is real.

15. Of course, this last situation, unlike the previous two, *does* occur frequently enough. But its occurrence or not depends on the particular case, while the underdetermination thesis is a *general* principle meant to apply to *all* cases.

16. In the famous paper in which Quine sets forth the modern version of the underdetermination thesis, he even allows himself to change the meanings of words and the rules of logic, in order to show that any statement can be held true, "come what may" (Quine 1980).

17. Or, as the physicist David Mermin calls them, "Duhem-Quine monstrosities" (Mermin 1998a).

2.2. Redefinitions of Truth

When facing the problems caused by underdetermination, one may be tempted by a radical turn: What about abandoning the notion of "truth" as "correspondence with reality," and seeking instead an alternative notion of truth? There are at least two currently fashionable proposals of this kind: one is to define truth through utility or convenience, the other is to define it through intersubjective agreement. The philosopher Richard Rorty offers examples of both:

> What people like Kuhn, Derrida and I believe is that it is pointless to ask whether there really are mountains or whether it is merely convenient for us to talk about mountains.[18] (Rorty 1998, 72)

> Philosophers on my side of the argument answer that objectivity is not a matter of corresponding to objects but a matter of getting together with other subjects—that there is nothing to objectivity except intersubjectivity. (Rorty 1998, 71–72)

Similar views are expressed by some of the founders of the Strong Programme in the sociology of science:

> The relativist, like everyone else, is under the necessity to sort out beliefs, accepting some and rejecting others. He will naturally have preferences and these will typically coincide with those of others in his locality. The words "true" and "false" provide the idiom in which those evaluations are expressed, and the words "rational" and "irrational" will have a similar function. (Barnes and Bloor 1981, 27)[19]

The best way to see that these redefinitions do not work is to apply them to simple concrete examples. For instance, it would certainly be useful to make people believe that if they drive drunk they will go to hell or die from cancer, but that would not make those statements true (at least on an intuitive understanding of the word "true"). Similarly, once upon a time, people agreed that the Earth was flat (or that blood was static, etc.), and we now know that they were wrong. So intersubjective agreement does not coincide with truth (again, understood intuitively).

Of course, we are using here an intuitive notion of truth, and a critic

18. See also the critiques by Nagel (1997, 28–30) and Albert (1998); and see Haack 1997 for an entertaining contrast between the two radically different "pragmatist" philosophies of C. S. Peirce and Rorty.

19. See Sokal and Bricmont 1998, chapter 4, for a critique.

might demand a more "rigorous" definition. But the problem is that all definitions tend to be circular or else to rely on fundamental undefined terms that one either grasps intuitively or does not grasp at all. And truth falls naturally in the latter category.[20]

Since these redefinitions of "truth" are so patently absurd, why are they proposed so often[21] and why are they so popular? Presumably, the answer has to do with the fact that, radical skepticism being irrefutable, one can always doubt any particular truth without running into logical contradiction. But these redefinitions do not even solve the problem of radical skepticism. Take, for instance, utility: saying that something is useful (for some specified goal) is already an objective statement (it has to be *really* useful for the declared goal) that relies implicitly on the correspondence notion of truth. The same remark is even more obvious for intersubjective agreement: to say that (other) people think so and so is an objective statement describing part of the (social) world "as it is."

Of course, positive arguments are sometimes given to support redefinitions of truth, as for instance the following somewhat subtle sophism:

> [T]he only criterion we have for applying the word "true" is justification and justification is always relative to an audience. So it is also relative to that audience's lights—the purpose that such an audience wants served and the situation in which it finds itself. (Rorty 1998, 4)

The beginning of the first sentence is correct, but it does not imply that truth is identical to justification. (One may well be rationally justified in believing something that turns out, on closer examination, to be false.[22]) Moreover, what does it mean to say that justification is always relative to the purpose that an audience wants served? This introduces a subtle confusion between knowledge and values, by implicitly assuming that all knowledge depends on some "purpose," that is, some noncognitive goal. But what if the "audience" wants to find out how (some part of) the world really is? Rorty might reply that this goal is unattainable, as

20. After all, people who ask what "truth" means are not really in the same position as those who wonder what an octopus is or who Xenophon was.

21. For a discussion of similar proposals, see Bertrand Russell's critique of the pragmatism of William James and John Dewey (Russell 1961, chaps. 24 and 25, in particular p. 779).

22. For example, Hume ([1748] 1988, sec. 10) gives the example of a person in India who, quite rationally, refused to believe that water can become solid during winter. (Water solidifies very abruptly around the freezing point, so if one lives in a warm climate, it is indeed hard to believe that water can freeze.) It shows that rational inferences from the available evidence do not necessarily lead to true conclusions.

the following statement suggests: "A goal is something you can know you are getting closer to, or farther away from. But there is no way to know our distance from the truth, not even whether we are closer to it than our ancestors were" (1998, 3–4). But is this really so? Some of our ancestors thought that the Earth was flat. Don't we know better? Aren't we closer to the truth, in that respect at least?

The view proposed here is so implausible that one is forced to resort to some "charitable" interpretation. Perhaps Rorty means by "truth" something like the fundamental physical laws governing the entire universe, or an "absolute" truth discovered by pure thought (as in classical metaphysics); and it does makes sense to be skeptical about our ability to discover truths of those kinds. But if this is what Rorty means, then he should say so explicitly, rather than making statements that allegedly apply to all possible knowledge. Or, alternatively, perhaps Rorty simply wants to reiterate the banal observation that all statements of fact (even about the flatness of the Earth) can be challenged by a consistent radical skeptic. But that is not a particularly new insight.

2.3 Then What Should One Do?

Given the problems raised by underdetermination, and since redefining truth leads us from bad to worse, what should one do? There is no abstract and general answer to this question. We are, in some sense, "screened" from reality (we have no immediate access to it, radical skepticism cannot be refuted, etc.). There are no absolutely secure foundations on which to base our knowledge. Nevertheless, we all assume implicitly that we can obtain some reasonably reliable knowledge of reality, at least in everyday life. Let us try to go farther, putting to work all the resources of our fallible and finite minds: observations, experiments, reasoning. And then let us see how far we can go. In fact, the most surprising thing, shown by the development of modern science, is how far we seem to be able to go.

A friend of ours once said, "I am a naïve realist. But I admit that knowledge is difficult." This is the root of the problem. Knowing how things really are is the goal of science; this goal is difficult to reach, but not impossible (at least for some parts of reality and to some degrees of approximation). If we change the goal—if, for example, we seek instead a consensus—then of course things become much easier; but as Bertrand Russell observed in a similar context, this has all the advantages of theft over honest toil.

It is important to remember that scientific knowledge needs no "justification" from the outside. The justification for the objective validity of scientific theories (in the sense of being at least approximate truths about the world) lies in specific theoretical and empirical arguments. Of course, philosophers, historians, or sociologists may be impressed by the successes of the natural sciences (as the logical positivists were) and seek to understand how science works. But there are two frequent mistakes to avoid: One is to think that, because some particular account fails (say, the logical-positivist one or the Popperian one), then some alternative account (e.g., the sociohistorical one) *must work*. But that is an obvious fallacy; perhaps *no* existing account works.[23] The second, and more fundamental, mistake is to think that our inability to account in general terms for the success of science somehow makes scientific knowledge less reliable or less objective. That confuses accounting and justifying. After all, Einstein and Darwin gave arguments for their theories, and those arguments were far from being all erroneous. Therefore, even if Carnap's and Popper's epistemologies were entirely misguided, that would not begin to cast doubt on relativity theory or evolution.

Moreover, the underdetermination thesis, far from undermining scientific objectivity, actually makes the success of science all the more remarkable. Indeed, what is difficult is not to find a story that "fits the data," but to find even one *noncrazy* such story. How does one know that it is noncrazy? A combination of factors: its predictive power, its explanatory value, its breadth and simplicity, etc. Nothing in the underdetermination thesis tells us how to find inequivalent theories with some or all of these properties. In fact, there are vast domains in physics, chemistry, and biology where there is a unique noncrazy theory that accounts for the known facts and where many alternative theories have been tried and failed because their predictions contradicted experiments. In those domains, one can reasonably think that our present-day theories are at least approximately true.[24]

23. See McGinn 1993, chapter 7, for the interesting suggestion that understanding our own knowledge-producing mechanisms simply lies outside the bound of what is biologically feasible for our limited minds.

24. With certain caveats to be made about the status of unobservable entities introduced in physics and the related debates between realists and instrumentalists. See Bricmont 2001 for a more detailed discussion. Note, however, that although relativists sometimes tend to fall back on instrumentalist positions when challenged, there is a profound difference between the two attitudes. Instrumentalists may want to claim either that we have no way of knowing whether "unobservable" theoretical entities really exist, or that their meaning is defined solely through measurable quantities; but this does not imply that they regard such entities as "subjective" in the sense that their meaning would be significantly influenced by extrascientific factors (such as the personality of the individual scientist or the social characteristics of the group to which she belongs). Indeed, instrumentalists may regard our scientific theories as, quite simply, the most satisfactory way that the human mind, with its inherent biological limitations, is capable of understanding the world.

Now that we have sketched our attitude on epistemological issues, let us turn to the consequences for contemporary sociological studies of science.

3. AGAINST RELATIVISM

There exist several contemporary variants of relativism. Quite often, people deny being relativist because they are relativist in some other sense than the one under consideration. So it is necessary to go over several possible meanings of the word. We shall consider here only two meanings: cognitive (or epistemic) relativism and methodological relativism. Our main thesis is that cognitive relativism is a position that no scientist (in either the natural or the social sciences) should wish to embrace, and that methodological relativism makes sense only if one adheres to cognitive relativism.

3.1 Cognitive Relativism

Roughly speaking, we will use the term "relativism" to refer to any philosophy that claims that the truth or falsity of a statement is relative to an individual or to a social group.[25]

The first thing to notice about cognitive relativism is that this doctrine follows naturally if we accept a radical redefinition of truth: clearly, if truth reduces to utility or to intersubjective agreement, then the "truth" of a proposition will depend on the individual or the social group in question. On the other hand, if we adopt the customary ("correspondence") notion of truth, then cognitive relativism is patently false: since a proposition is true to the extent that it reflects (some aspects of) the way the world is, its truth or falsity depends on the way the world is and not on the beliefs or other characteristics of any individual or group.

Since we have already discussed redefinitions of truth, there is not much to add, except that it makes no sense for ordinary scientists—whether they study nature or society—to adopt, even implicitly, a cognitive relativist attitude, for cognitive relativism amounts to abandoning the goal of objective knowledge pursued by science. However, it seems that some historians and sociologists want to have it both ways: adopt a relativist attitude with respect to the natural sciences and an objectivist

25. We will consider only relativism about statements of fact (i.e., about what exists or is claimed to exist) and leave aside relativism about ethical or aesthetic judgments.

(even naïve realist) attitude with respect to the social sciences.[26] But that is inconsistent; after all, research in history, and in particular in the history of science, employs methods that are not radically different from those used in the natural sciences: studying documents, drawing the most rational inferences, making inductions based on the available data, and so forth. If arguments of this type in physics or biology did not allow us to arrive at reasonably reliable conclusions, what reason would there be to trust them in history or sociology? Why speak in a realist mode about historical categories, such as Kuhnian paradigms, if it is an illusion to speak in a realist mode about scientific concepts (which are in fact much more precisely defined) such as electrons or DNA?

3.2 Methodological Relativism

Methodological relativism is not in itself a philosophical position; it is rather, as the name indicates, a set of methodological principles. This relativism is associated to developments in the history and sociology of science that began during the 1970s under the banner of the so-called Strong Programme and which have had an enormous impact in the field of sociology of scientific knowledge (SSK) as well as outside that field (in cultural studies, anthropology, etc.).[27] The Strong Programme proposes to give a causal account of the acceptance of scientific ideas, while remaining "impartial" (or "symmetrical") as to whether they are true or false, rational or irrational. Here is how David Bloor lays out the principles for the new sociology of knowledge [22]:

> 1. It would be causal, that is, concerned with the conditions which bring about belief or states of knowledge. Naturally there will be other types of causes apart from social ones which will cooperate in bringing about belief.
>
> 2. It would be impartial with respect to truth and falsity, rationality or irrationality, success or failure. Both sides of these dichotomies will require explanation.
>
> 3. It would be symmetrical in its style of explanation. The same types of cause would explain, say, true and false beliefs.
>
> 4. It would be reflexive. In principle its patterns of explanation would have to be applicable to sociology itself. (1991, 7)

26. See Sokal and Bricmont 1998, chapter 4, for relevant quotes from Kuhn, Feyerabend, Barnes-Bloor, and Fourez, along with a more detailed critique.

27. See Laudan 1981, 1990a, Slezak 1994a, 1994b, and Kitcher 1998 for related criticisms of the Strong Programme. Note particularly Kitcher's criticism of the "Four Dogmas of Science Studies" (38–45), which is quite similar to our own critique.

How is one to understand the symmetry and the impartiality theses? In order to see the difficulty, let us first consider perception in everyday life (we'll turn to scientific theories in a moment). Suppose that several of us are standing outdoors in the rain, and someone says, "It is raining today." That statement expresses a belief; how are we to explain this belief "causally"? Well, no one today knows the complete details of the causal mechanisms, but it seems obvious that part of the explanation involves the fact that it really is raining today. If someone said that it is raining when it is not, one might think that he is joking or that he is mentally disturbed, but the explanations would be very asymmetrical, depending on whether it is raining or not.[28]

Faced with this problem, supporters of the Strong Programme could admit what we say for ordinary knowledge, but maintain that it does not apply to scientific knowledge: in the latter, reality would play little or no role in constraining our beliefs.[29] However, this claim looks particularly implausible, since scientific activity—far more so than everyday life—is set up (through experiments, etc.) precisely so as to make Nature itself constrain our beliefs about it as strongly as possible [18].

Let's consider, once again, a concrete example: Why did the European scientific community become convinced of the truth of Newtonian mechanics sometime between 1700 and 1750? Undoubtedly a variety of historical, sociological, ideological, and political factors must play a part in this explanation—one must explain, for example, why Newtonian mechanics was accepted quickly in England but more slowly in France[30]—but certainly *some* part of the explanation (and a rather important part at that) must be that the planets and comets really do move (to a very high degree of approximation, though not exactly) as predicted by Newtonian mechanics.[31]

28. See Gross and Levitt 1994, pages 57–58, for a similar discussion. Of course, even ordinary perception is "social" in some sense. For example, in order to see clearly, some people need eyeglasses that are socially produced. More fundamentally, the meaning of the words through which one expresses one's perceptions is to some extent influenced by the environment in which they are used. Sometimes relativists insist that all they claim is that science is "social" in some equally weak sense, but that seems to us like a considerable watering-down of the "symmetry" thesis. Indeed, when one studies perception scientifically, there is no "symmetry," in any meaningful sense, between hallucination and correct perception. And the difference between the two is related to how the world really is, so that the latter is partly causally responsible for correct perceptions.

29. See note 1 above for an explicit assertion of this thesis by Harry Collins.

30. See, for example, Brunet [1931] 1970 and Dobbs and Jacob 1995.

31. Or more precisely: There is a vast body of extremely convincing astronomical evidence in support of the belief that the planets and comets do move (to a very high degree of approximation, though not exactly) as predicted by Newtonian mechanics; and *if* this belief is correct, then it is the fact of this motion (and not merely our belief in it) that

At the risk of beating a dead horse, let us rephrase our critique of the Strong Programme's sociological reductionism as a *reductio ad absurdum*. Consider the following thought experiment: Suppose that a Laplacian demon were to give us all conceivable information about seventeenth-century England that could in any way be called sociological or psychological: all the conflicts between members of the Royal Society, all the data about economic production and class relations, etc. Let's even include documents that have been destroyed and private conversations that were never recorded. Add to this a gigantic super-fast computer that can process all this information as much as desired. But do not include any astronomical data (such as Kepler's observations). Now, try to "predict" from those data that scientists will accept a theory in which the gravitational force decays with the inverse square of the distance, rather than the inverse cube. How could one do it? What kind of reasoning could one use? It seems obvious to us that this result simply cannot be "extracted" from the given data.[32]

Now suppose, by contrast, that one wants to give a causal account of belief in astrology. In this case it is at least conceivable that one could obtain a purely sociological or psychological account of the incidence of such beliefs, without ever invoking the good evidence supporting those beliefs—simply because there is no such evidence.[33] This comparison between Newtonian mechanics and astrology shows clearly a necessary and crucial asymmetry in the explanatory scheme: in the one case, evidence must enter into any satisfactory explanation, in the other case, not. Note, of course, that if you happen to believe (wrongly) that astrology *is* well supported by evidence, then this factor *should* presumably enter into what you regard as a satisfactory causal account of belief in astrology.

In summary, it seems clear that an adequate causal explanation of how scientific theories come to be accepted would have to combine "natural" and "social" factors, just as for ordinary perception. Of course, explaining scientific knowledge is much more complicated than explaining perception, which is complicated enough [15].

---

forms part of the explanation of why the eighteenth-century European scientific community came to believe in the truth of Newtonian mechanics.

32. Of course, one can argue that the rise of science is linked to the rise of the bourgeoisie (although the causal link between the two, if any, is unclear); one might even argue that a "mechanical worldview" is associated with the bourgeois ethos. But that kind of argument will not extend to detailed empirical statements like the inverse-square law.

33. Of course, one may have a separate worry: Does anyone at present have a well-tested sociological or psychological theory that yields a causal and explanatory account of *any* system of beliefs, even superstitious ones?

Earlier in this essay, we made an analogy between scientific investigations and police inquiries. Continuing this analogy, one could say that ontological relativism amounts to saying that there is no objective fact of the matter about whether a particular suspect is innocent or guilty, while epistemological relativism is the assertion that no method of inquiry can be said to be objectively better than another (e.g., carefully analyzing fingerprints versus planting evidence). Methodological relativism, on the other hand, amounts to trying to understand how the police, judge, and jury become convinced of X's guilt without ever taking into account the fact that, in some cases at least, there might be good evidence for X's guilt.

Let us consider in this light an assertion of Collins and Pinch about Einstein's theory of relativity:

> Relativity . . . is a truth which came into being as a result of decisions about how we should live our scientific lives, and how we should licence our scientific observations; it was a truth brought about by agreement to agree about new things. It was not a truth forced on us by the inexorable logic of a set of crucial experiments. (1993, 54)

Wouldn't it sound odd to say that "it is true that X is guilty" but that this truth "came into being as a result of decisions about how we should licence our police investigations; it was a truth brought about by agreement to agree about new things" [18]? The whole thing is plagued with ambiguities: Does one mean to say that X is guilty or not? Is this merely a confusing way of stating the banal observation that our *belief* in X's guilt arose from a social process?[34]

When all is said and done, methodological relativism makes no sense unless one adheres to the idea that the natural sciences form some kind of ideology or religion, while our knowledge of the social world is truly scientific and explains (or will someday explain) why natural scientists believe what they do. But then, we have a direct competition: Which theories are more scientific, that is, are better supported by evidence,

34. Let us note in passing that the last sentence of the quote is correct: the notion of "crucial experiment," which is used by some philosophers of science, grossly oversimplifies the complex web of interlocking evidence that gives support to well-confirmed scientific theories. The physicist David Mermin, in his excellent critique of the account of relativity given by Collins and Pinch, correctly concedes that scientists' oversimplified histories, as presented in textbooks, sometimes do make this error (Mermin 1996a, 1996b, 1996c, 1997). On the other hand, experiments and observations, *taken collectively,* are indeed crucial since there is no other way to obtain reliable knowledge of the external world.

make more accurate predictions, etc.? Those of physics and chemistry and biology, or those of sociology (including the sociology of religion and of fashion)? The answer seems clear enough.[35,36] This unpleasant situation (for sociologists of science) sometimes leads them to employ arguments supporting cognitive relativism, which have the "merit" (from their point of view) of stopping the "direct competition": if no theory is objectively better than another, then physics is not more scientific than sociology. But, as we explained previously, cognitive relativism is not a view that any scientist—natural or social—should want to hold.[37]

It is interesting to compare SSK and postmodernism, which are often confused by their opponents. Postmodernists tend to reject objectivity even as a goal: everything becomes dependent on one's subjective viewpoint, and moral or aesthetic values displace cognitive ones. Quite the opposite is true for supporters of the Strong Programme, who often appear extraordinarily scientistic: for example, Bloor often emphasizes that his view is materialistic, naturalistic, and scientific. But the methodological relativism contained in the "symmetry" and "impartiality" theses—unless it is interpreted in so watered-down a fashion as to become virtually empty—undermines the rationalist aspect of the sociological enterprise, so that practitioners of SSK eventually have to fall back on radical-skeptical arguments about objectivity that often lead them to make common cause with the postmodernists.

35. To avoid misunderstandings, let us emphasize that this does not mean that natural scientists are more clever than sociologists or historians, but simply that they deal with easier problems.

36. In a similar vein, the chapter of Barnes, Bloor, and Henry 1996 on "proof and self-evidence" is eerily fascinating. The authors try to refute the claim that some beliefs, like $2 + 2 = 4$ or the *modus ponens,* are so obvious that they need not be explained sociologically. But their arguments show, at most, that those beliefs are not as evident as they may seem (e.g., because the nature of arithmetic statements is open to divergent interpretations in the philosophy of mathematics, or because the *modus ponens* applies only to ideally precise propositions and not to those containing ill-defined words like "heap"). But that answer misses the obvious point that all human beings—be they physicists or sociologists or plumbers—have, in practice, no sensible alternative but to use arithmetic and logic. And to seek a sociological explanation for such basic notions surely puts the cart before the horse. Do Barnes, Bloor, and Henry really think that their sociological theories are more reliable than $2 + 2 = 4$ and *modus ponens*? See Nagel 1997 for an elaboration of these arguments, and Mermin 1998a for another critique.

37. For charity, we have here left aside Bloor's fourth principle ("reflexivity"). Indeed, it seems to us that if sociologists start trying to explain why they hold their own beliefs without taking into account the evidence that those beliefs are somehow better or more objective than those of their critics, then we simply move from error to absurdity. Note that, by contrast, Collins (1992, 188) argues that "sociologists of scientific knowledge who want to find (or help construct) new objects in the world must compartmentalize; they must not apply their methods to themselves." That move allows him to escape from self-refutation, but why should anyone accept his rule? See Friedman 1998 for a more detailed discussion.

4. CONCLUSION: REAL ISSUES

We do not want to give the impression that there are no interesting questions to be dealt with by sociologists of science. On the contrary, there are lots of them. But we contend that the philosophical confusions currently fashionable in SSK circles hinder rather than foster the possibility of seriously studying them.

The kind of questions we have in mind revolve primarily around the problem of expertise. We are constantly subjected to reports of "expert" opinion, on all possible topics. But should one believe them? Should one believe that tobacco is bad for one's health? That olive oil is good (after all)? That nuclear power plants are safe? That the austerity measures of the IMF are good for the economy? That newspaper reports are accurate? That publications on the "memory of water"[38] concern some real physical effect?

When confronted with experts, any individual or small group of individuals is in a difficult situation. There is no way to find the time and the means to check directly even a small fraction of the experts' assertions. And yet, in many practical situations we have to decide whether or not to trust their claims. How should we proceed? That is a truly interesting and difficult question. But epistemic and methodological relativism do not help here. We want to find out who is right and who is wrong, and that depends ultimately on how the world really is [17]. Nor is the question particularly new: for example, Hume addressed it already, and gave some guidelines for solving it, in his discussion of whether one should believe in miracles ([1748] 1988, sec. X). The argument is well known: If you have never seen a miracle yourself, your belief is based on believing someone who reports the occurrence of a miracle. But you know from direct experience that people sometimes deceive themselves or cheat others. So, whenever you hear the report of a miracle, it is always more rational to believe—at least in the absence of powerful countervailing evidence—that some kind of deception is taking place, rather than a true miracle [15].[39]

To give a concrete example of the kind of reasoning that we have in mind, consider the issue of the "memory of water." One way to make a "Humean" argument about its plausibility goes as follows: Given that

38. An effect allegedly found by French scientist Jacques Benveniste that, if true, would give theoretical support to homeopathy (Davenas et al. 1988). See Maddox, Randi, and Stewart 1988 for a critical analysis and, for a more detailed discussion, see Broch 1992.

39. As Hume observed, it is even rational for an inhabitant of India to disbelieve the claim that water can become solid during winter: see note 22 above.

the result, if true, would provoke a revolution in physics and chemistry, at least some scientists around the world should have an interest (in both senses of the word) in duplicating the result. Moreover, the experiment itself does not require huge investments. However, no replication has been claimed, at least not by people totally independent of Benveniste.[40] Negative results are usually not reported, so skepticism with respect to the original experiment becomes reasonable (to say the least). Of course, we are here only suggesting the outline of an argument. A real investigation would have to find out whether the experiment really was easy to replicate and whether attempts at replication (leading to negative results) were in fact made. And that involves considerations both of physics and of sociology.

However, one of the disturbing aspects of SSK's methodological relativism is that its adherents tend to combine an exaggerated skepticism towards conventional scientific knowledge with a rather tolerant (or even favorable) attitude toward the pseudosciences [20, 21]. For example, Barnes, Bloor, and Henry comment sympathetically on homeopathy and astrology (1996, chap. 6) and go so far as to assert that the data gathered by Gauquelin in support of the astrological theory that there is a "Mars effect" affecting the destiny of sports champions "could conceivably come to be accommodated as a triumph of the scientific method" (p. 141). As Mermin observes, following a Humean type of reasoning, the data of Gauquelin are so surprising (amounting to a "miracle") that it is reasonable to assume rather that some kind of deception (or self-deception) has occurred (Mermin 1998a).[41] Another example is provided by the comment by Collins and Pinch that "if homeopathy cannot be demonstrated experimentally, it is up to scientists, who know the risks of frontier research, to show why" (1993, 144). But this amounts to shifting the burden of evidence: it is up to advocates of homeopathy to "demonstrate experimentally" that their therapy works beyond the placebo effect, not the other way around [15, 21]. It is unfortunately easy to find similar statements: for instance, in one of the early books on "Alternatives to Big Science," one could already read: "future prospects for a great breakthrough in Western science and in the occult sciences of the Orient are said to be good. Astrology will again become a recognised science, once it has made use of cybernetics and statisti-

40. Unlike, for example, the experiments showing the existence of superconductivity at high temperatures, which were replicated around the world within a few weeks.

41. See also Benski et al. 1996 for a critical and detailed factual examination of the "Mars effect."

cal analysis. But such claims fall on the deaf ears of official science" (Nowotny 1979, 15).

These remarks are shocking but are perhaps not so surprising, since the "neutrality" of the Strong Programme leads its adherents to disregard any epistemological distinctions between science and pseudoscience.

Of course, believing experts is not the same thing as believing in miracles (or in the pseudosciences); and in order to formulate deeper principles of rational inference in real-life situations, many sociological considerations become relevant. First of all, it is important to have at least a rough idea of how "experts" in a given field become accredited and what methods they use within their area of "expertise": this allows one to distinguish, on epistemological grounds, between licensed medical doctors and equally highly "accredited" practitioners of aura reading. Then, roughly speaking, one should believe a (genuine) expert when there are other, equally competent experts that have both an interest and the means to contradict the expert in question and do not do so. But that involves numerous sociological questions: How free is the so-called free market of ideas? Do contradictory viewpoints get a fair hearing? Nothing prevents the existence of what one might call "democratic Lysenkos," namely people who, within democratic societies, get hold of some institutional position of power (a scientific journal, a research institute) and impose their favorite "line" of research there, leading to a dead end. In fact, anyone working in a university knows that there are lots of democratic Lysenkos, at least on a small enough scale. A most interesting problem for sociologists (and policymakers) is to design institutions that minimize the likelihood that such people become too powerful.

When all is said and done, there is no need for a "science war" between scientists and sociologists. Both could perfectly well cooperate on a variety of issues. In our view, science studies' epistemological and methodological conceits are a diversion from the important matters that motivated science studies in the first place: namely, the social, economic, and political roles of science and technology. To be sure, those conceits are not an accident; they have a history, which can itself be subjected to sociological study.[42] But science studies practitioners are not obliged to persist in a misguided epistemology; they can give it up and go on to the serious task of studying science. Perhaps, from the perspec-

42. For an interesting conjecture, see Nanda 1997, 79–80. For a different (but not incompatible) conjecture, see Gross and Levitt 1994, 74, 82–88, 217–33. Both these conjectures merit careful empirical investigation by intellectual historians.

tive of a few years from now, today's so-called science wars will turn out to have marked such a turning point.

ACKNOWLEDGMENTS

We would like to thank Michel Ghins, Shelley Goldstein, Antti Kupiainen, Norm Levitt, and Tim Maudlin for many interesting discussions on the issues addressed here. Of course, they are in no way responsible for what we have written.

Chapter 4

IS A SCIENCE PEACE PROCESS NECESSARY?

Michael Lynch

Just keep talking. —Stephen Hawking

Rumor has it that "science wars" have been going on between natural scientists and sociologists.[1] This rumor spread in the aftermath of physicist Alan Sokal's (1996b) hoax of the cultural studies journal *Social Text*. Sokal submitted an article that praised the progressive tendencies of left-leaning critics of positivism and asserted that cutting-edge developments in quantum theory resonate with poststructuralist literary theory. As soon as the article was published, Sokal (1996a) announced that he had written such nonsense in order to expose the widespread ignorance and tendentiousness in the science studies and cultural studies fields. Ironically, the bogus article appeared in a special issue of *Social Text* entitled "The Science Wars," which was put together in order to combat the charge that social and cultural studies of science were exercises in science bashing. The clamor about Sokal's hoax completely drowned out this message.

As usually portrayed, the science wars involve a conflict between two opposing camps: natural scientists and sociologists. The sociologists are identified with the far left of the political spectrum, while the scientists, or at least their most significant sponsors and cheerleaders, are associated with the right. The scientists are said to believe in nature, truth, and reality, while the sociologists are said to believe that representations

1. For an amusing etymology of the "science wars," see Labinger 1997, 203. The derivation from George Lucas's *Star Wars* anthology may account for the use of the plural form for what otherwise may seem to be a singular "science war." A more immediate point of reference is the "culture wars": a series of controversies over university "great works" curricula and museum exhibitions.

of nature are arbitrary, scientific laws are ideological, and "reality" is a myth. Although words are the main weapons, this dispute is likened to a war in which the differences between the two sides are so deep that there is little hope of appealing to common ground in order to reach rational agreement. In line with this image, the partisans often seem less interested in conducting a debate than in convincing a larger audience that the opposition's views are nonsensical, insincere, and potentially dangerous. Even, and perhaps especially, some of the defenders of science and reason heap ridicule upon their opponents. Hyperbole and ad hominem argument are commonplace, along with charges and countercharges of dishonesty. Richard Dawkins (1998, 143) credits Sokal with a "brilliant hoax" that exposed a whole host of *impostors* in the humanities and social sciences, whereas Stanley Fish (1996) suggests that Sokal perpetrated a *fraud.*[2] Another writer (a philosopher) goes so far as to counsel intolerance for the "intellectual charlatanism" that pervades broad areas of twentieth century philosophy and social research (Bunge 1996). Although such charges are made by self-declared defenders of "truth and reason," by opting for crude assertions, name-calling, and ridicule, instead of dialogue and argument, they undermine what they would defend.[3] Though falling short of full-scale war, the form of dispute is less like an academic debate and more like an adversary lawsuit, but without the formal restrictions on argument and evidence.

Many of us are unhappy with the polarization associated with the "science wars" and would like to promote a "science peace process." Before initiating such a peace process, however, it may be worth considering whether war is going on in the first place. If the "war" analogy is inappropriate in this instance, then there should be no need to declare peace. A peace process is an effort to start "talks" aiming to resolve, or at least interrupt, a chronic and violent conflict. The idea is that the combatants should talk instead of trying to destroy one another. Contrary to the usual kind of war, the science wars have been *verbal* debates between members of different academic fields. Until recently, many of the debaters had little to say to one another. As arguments often do, these particular debates frequently become heated and denunciatory, but it seems unnecessary to suspend a state of war in order to initiate peace talks. The state of *this* war consists of talk, and more talk, so it

2. Dawkins (1998, 141) confidently equates the obscure term "impostures" in Sokal and Bricmont's (1998) title with the more familiar "impostors."

3. I can speak on behalf of the accused charlatans because Bunge (1996, 100) makes a brief, crude, and inaccurate characterization of one of my articles (Lynch 1988).

may make more sense to devise more interesting and respectful ways of talking and writing.

In this essay, I will briefly outline why I think the science wars analogy is misleading, and then I will discuss two recurrent arguments that have featured prominently in the debates. I will close with some suggestions about how those of us who are interested in doing so might turn the science wars dispute into an educational opportunity.

DEPOLARIZING THE "WARS"

It is simple and dramatic to present the science wars as a conflict between two clearly defined sides. Frequently, one side is given the name "scientists" and the other, "sociologists." This division sometimes is mapped onto C. P. Snow's (1959) distinction between the two cultures of the sciences and humanities. Further dichotomies between left- and right-wing political ideology and realist and antirealist metaphysics complete the picture of a grand conflict. A concise example of such a characterization is an anonymous editorial in *Physics World* (June 1997, 3) entitled "Reality Is Not a Hoax."[4] The essay begins by saying that the "'Sokal hoax' has become a mini-industry—a by-word for the increasing friction at the interface between the natural and social sciences." The editorial then goes on to say that "[t]he fact that Sokal's paper was published does not in itself prove anything profound. However, it did drag into the open an on-going squabble between a group of scientists, many of them physicists, and various sociologists of science. Many scientists feel uneasy about various ideas from sociology, notably the suggestion that the laws of nature as we know them are 'social constructs'—essentially laws that scientists have agreed between themselves—and do not have any fundamental significance. 'Relativism' is another name for this school of thought." The essay was written in a spirit of reconciliation, and the writer goes on to suggest that each side has something to learn from the other. While I support the overall recommendation, I have some doubts about the initial portrait of an "ongoing squabble" between two clearly identified sides representing divergent "schools of thought."

The Two Sides

The scientists are generally portrayed as a more unified group than their opponents, although their side also includes many camp followers from

4. I am grateful to David Edge for having sent me a copy of the editorial, along with many other related articles in popular and scientific sources.

the humanities and social sciences. For example, of the many attacks on relativism, cultural constructionism, and postmodernism in Gross, Levitt, and Lewis's *Flight from Science and Reason* (1996), some of the most "unremittingly negative" (to borrow philosopher Susan Haack's [1996, 264] characterization of her own contribution) are written by philosophers and social scientists (Bunge 1996; Cole 1996). Gerald Holton (1993), a physicist who is much better known for his contributions to history and sociology of science, has been at the forefront of the "science" side of the fray, and the editors of the journal *Philosophy and Literature* were so pleased with Alan Sokal's "hoax" article that they invited him to serve on to their editorial board (Dutton and Henry 1996).

Even among bona fide scientists, there is some ambiguity about what "school of thought" they support and how strongly they support it. Paul Gross and Norman Levitt (1994) have been the most prominent spokesmen for the "science" side, but others have distanced themselves from their views. Richard Lewontin (1998, 59), for example, begins an essay by repudiating Gross and Levitt's "charmingly naive view of science." Lewontin (1996) and biologist Ruth Hubbard (1996) are also included in the expanded book version of the "Science Wars" special issue (Ross 1996a). It is also the case that several of the "sociologists" whose writings have been criticized as antiscientific, as ignorant of science, or both, hold advanced degrees and have research experience in physics, biology, astronomy, and engineering. In addition, many critical and controversial remarks reported by ethnographers of science are presented in the form of quotations from scientist-informants (see, for example, many of the quotations in Gilbert and Mulkay's [1984] study of a scientific controversy). While some quotations may be idle remarks taken out of context, others, like the lengthy quotations from Tom White and other molecular biologists in Paul Rabinow's *Making PCR* (1996), carefully and explicitly discuss dismaying trends and dubious conduct in the natural sciences. So Gross and Levitt do not necessarily speak for a unified front.

It is more widely recognized that the "sociology" side (composed of alleged antiscientists, relativists, cultural constructionists, postmodernists) is diffuse and divided. "Sociologists" is an inaccurate label for a group that includes historians, philosophers, anthropologists, and other specialists. The frequently made associations between sociology, relativism, and postmodernism are also dubious. Comparatively few professional sociologists work in the subfield of sociology of science, and many who do are not "relativists." Even many "relativist" sociologists of science subscribe to positions that they argue are compatible with empirical science (Bloor 1991; Barnes, Bloor, and Henry 1996; Collins 1983) and incompatible with "postmodernism" and even "social constructionism" (Latour 1990; 1999b).

The Political Alignments

The "sociology" side is generally identified with radical leftist politics, and some writers argue that student radicals from the 1960s have grown up to become professors and academic administrators who dominate particular departments and colleges. So, at a time when radical socialism has lost much of its hold even in nominally socialist parties, it thrives in the humanities and social sciences. Some spokespersons for the "sociologists" accuse the "scientists" of being right-wingers, or if not right-wingers, unwitting mouthpieces for right-wing sponsors (Ross 1996b, 7). The accused mouthpieces deny the charges, and some of them (like Gross, Levitt, and Sokal) profess sympathies for the political left, while arguing that the absurd beliefs promulgated by the "academic left" actually undermine genuine leftist hopes for political change. Similar arguments have been made by writers in politically left periodicals (e.g., Frank 1996). Countercharges of "left-conservatism" have done little to clarify the waters. Whether or not one is cheered or depressed by such internal squabbles and ambiguities, they indicate how difficult it can be to assign the various factions in the science wars to clear-cut political positions on the left or right.

The Metaphysical Positions

While it may be difficult to define the two sides that are supposedly at war, everyone seems to agree that the conflict involves two diametrically opposed positions. But this agreement raises its own set of difficulties. Nobody disputes the fact that many scientists, including some of the most distinguished scientists alive today, have voiced their disapproval of social and cultural studies of science. However, it is no less obvious that the books and articles in which these scientists have registered their views differ remarkably from their more numerous publications in specialized journals. A reader who knows nothing about the background of the authors who wrote *Intellectual Impostures* (Sokal and Bricmont 1998), *The Unnatural Nature of Science* (Wolpert 1992), or *Higher Superstition* (Gross and Levitt 1994) will learn that Sokal and Bricmont are physicists, Wolpert is an embryologist, Gross is a retired biologist, and Levitt is a mathematician, but the reader will learn nothing about the authors' research. There is no deception in this, because the science wars polemics are not written as specialized contributions to physics, biology, or mathematics, even when they are published as editorials and letters in journals like *Science* and *Nature*. The polemics are written for a broad audience including readers who have limited training in the natural sciences. The authors discuss general concepts of truth and rationality, they attack relativism, they discuss the relationships between science and common sense, and they speculate about the nature of reality. Gross, Levitt, Sokal, Wolpert, Weinberg, and their scientist colleagues do not claim to have backgrounds in philosophy, and some of them profess indifference to philosophy, but their arguments tend to be philosophical in scope. While they draw upon knowledge of science and mathematics, they also make abundant use of historical examples and cultural analogies. Most of their "relativist" opponents are not philosophers (or, for that matter, philosophical relativists), so the debate is, in many respects, an exercise in pop metaphysics in which professional philosophers only occasionally lend a hand. I do not mean to imply that the debate is worthless, or that the participants have no business engaging in it. At least according to one traditional view, philosophical questions are open to everyone, and it also can be argued that practicing natural scientists and sociologists who have investigated scientific practices know more about science than they would learn by mastering the p's and q's of technical philosophy. However, it is worth keeping in mind that an expertise in quantum physics or embryology does not necessarily qualify a scientist to address general questions about the relationship between scientific theories and the natural world. Nor is it the case that empirical studies in the sociology of science necessarily provide clear and certain understandings of the nature of science, knowledge, and truth.

The science wars are thus a very strange conflict. The science side includes many nonscientists, and at least a few natural scientists have distanced themselves from the more vigorous spokesmen for "their" side. The sociology side has little clear connection with the professional discipline that goes by that name, and its affiliation with "left" political causes is ambiguous and internally contested. Although the debate is ostensibly about science, the content of the debate is philosophical. Contrary to the typical way the science wars are portrayed, the conflict is a metaphysical battle fought by conscripts who have limited training in the martial arts of philosophy. A few philosophers have weighed in on each side, but for the most part the metaphysical arguments have been carried on by researchers, scholars, and journalists who are more at home in other fields.

The science wars are open to all comers. It helps to be a professional scientist (preferably a famous one) if you want to represent the science side, and it helps to have conducted empirical case studies of scientific practice if you want to speak up for the sociologists. Training in philosophy is a bonus, but it isn't really necessary for playing the game. The entry requirements are low, and it appears that the whole world is joining in. Philosopher Ian Hacking reports that "one of the search engines on the internet found 84,272 distinct items in which 'Sokal Affair' was a key phrase. For comparison, it found 7767 for Ludwig Wittgenstein and 11,334 for Quantum Mechanics, the science whose abuse furnished Sokal's illustrations" (1997, 14). There is at least one Web site devoted exclusively to the science wars, and perhaps soon there will be university courses and even entire departments of science war studies. Many of the more visible participants in the debates have impressive technical credentials, and sometimes they call upon their specialized backgrounds when giving examples or criticizing errors in their opponents' arguments, but the most recurrent arguments and many of the standard examples in the debate are not at all technical. Prospects for peace are inherent in this situation. It is not a question of initiating talks—we are already drowning in talk and writing—but of "elevating" the existing level of argument so that the science wars become more of an opportunity for the countless participants to learn something valuable from this vast exchange.

Two Arguments

Two arguments that crop up again and again in the polemical exchanges are ripe for upgrading. I call them the Airplane Argument and the Baseball Argument. These are not technical arguments, and joining them does not require scientific training or philosophical sophistication. The Airplane Argument is the simplest and most basic of the two, so I will start with it before going to the Baseball Argument.

The Airplane Argument

A concise summary of this argument is given by Richard Dawkins (1994), who says that constructionists refute their own arguments every time they step aboard commercial aircraft—something the better-known constructionists do regularly on their way to international conferences. In a variant of this argument, Sokal raises the following challenge: "Anyone who believes the laws of physics are mere social conventions is invited to try transgressing those conventions from the windows of my apartment. (I live on the twenty-first floor)" (1996a, 62). Such "refutations" are frequently made, but it is puzzling that they are taken seriously. The same social constructionists who Dawkins implies should be fearful of flying, should, according to Sokal, be unconcerned about the consequences of falling from high buildings. Sokal states that any sane person should know better than to leap from his window. This is true as a matter of common sense, but it is unclear what it has to do with the laws of physics. Even people who incorrectly understand the laws of physics are likely to know better than to take up Sokal's challenge. Galileo's Aristotelian critics are now said to have held mistaken conceptions of motion. Many of them also believed in angels, and yet they apparently knew better than to leap off of the Tower of Pisa.

The airplane argument equates social constructionism with a profound distrust in the reliability of gravitational forces and engineered systems, and Sokal's challenge goes even further by equating constructionism with a lack of common sense. We are left with a puzzle about what is meant by the "laws of physics." Are we supposed to equate the laws of gravity with our everyday knowledge of how things fall to the ground, or should we equate them with the technical systems devised by Newton, Einstein, and other physicists? If someone believed that gravity is haphazard, flying in airplanes would be the least of their worries, and whether or not we believe that a regularity is "natural" or "constructed" has little to do with whether or not we trust it. Anyone who steps aboard a commercial airliner must trust a complex social system of aircraft design and construction, pilot training, aircraft maintenance, and air-traffic control. Material regularities are crucial aspects of that system, but they are far from the whole story. I have known aeronautics engineers who said they were reluctant to step aboard DC-10 airliners because of a history design flaws resulting from the corporation's frantic pursuit of a contract. They did not distrust the aircraft because it was constructed, but because of the *way* it was constructed. A fear of flying, or the absence of fear of falling, has no clear relation to a general belief in the constructed or unconstructed nature of reality.

Sokal's example suggests that belief in the laws of physics is a matter of common sense. Elsewhere, Sokal and Bricmont (1998, 55, note 56) argue for continuity between scientific theories and commonsense knowledge of physical regularities. Lewis Wolpert, another scientist who criticizes social constructionism, makes the opposite argument (1992), that scientific knowledge is "unnatural," or counterintuitive, insofar as it requires specialized training and ways of thinking that are beyond most people's capacities. So we are left with two potentially interesting questions: What is meant by a physical law? and What is the relationship between laws of physics and commonsense knowledge (and trust) of material regularities?

The Baseball Argument

Stanley Fish started the ball rolling, so to speak, in a *New York Times* essay (1996) in which he defended social constructionism against the wave of popular ridicule that crested in the days following Sokal's announcement of his hoax (1996a). Fish argued that it is a mistake to assume that when sociologists say that a fact is "socially constructed," they are claiming that it is "not real." He used an analogy with baseball to illustrate this point: "[B]alls and strikes are both socially constructed and real, socially constructed and consequential. The facts about balls and strikes are also real but they can change, as they would, for example, if baseball's rule makers were to vote tomorrow that from now on it's four strikes and you're out." Fish acknowledged that scientists do not establish the laws of physics by voting. It would be more in line with his suggestion to suppose that they vote with their feet. Scientists cast votes, of sorts, when they cite prior research and selectively incorporate others' techniques and results into their own research. And, the term "referee," used for reviewers of journal articles and grant proposals, suggests analogies with judgments about rules in field games (though not baseball, where umpires preside). Fish observed that scientists do not "present their competing accounts to nature and receive from her an immediate and legible verdict" (1996, A23). The verdict depends upon judgments made by historically and institutionally situated agents. Steven Weinberg objected to this argument, and said that the reality of physical laws is more like that of "the rocks in the field" than the rules of baseball:

> We did not create the laws of physics or the rocks in the field, and we sometimes unhappily find that we have been wrong about them, as when we stub our toe on an unnoticed rock, or when we find we have made a mistake (as most physicists have) about some physical law. But the languages in which we describe rocks or in which we state physical laws are certainly created socially, so I am making an implicit assumption (which in everyday life we all make about rocks) that our statements about the laws of physics are in a one-to-one correspondence with aspects of objective reality. To

> put it another way, if we ever discover intelligent creatures on some distant planet and translate their scientific works, we will find that we and they have discovered the same laws. (1996, 13)

Weinberg's mention of toe-stubbing rocks borrows from Dr. Johnson's classic refutation of idealism. Weinberg's explicitly asserted "implicit assumption . . . that our statements about the laws of physics are in one-to-one correspondence with aspects of objective reality" also is a familiar enough philosophical position. It is a position that has been debated among philosophers for a long time. Weinberg's claim about intelligent creatures from distant planets is more novel, and also more dubious: until a few centuries ago, the "scientific works" of the intelligent creatures who inhabited *Earth* did not correctly describe the laws of physics that are accepted today.

The laws of physics and their constituent concepts of force, mass, space-time, energy, etc., are historical legacies. The same can be said about general concepts like "fact," "law," and "discovery" and the distinction between objectivity and subjectivity. A number of historians have noted that objectivity and subjectivity are concepts that have changed definition over the past few centuries. Many philosophers—including some hard-headed logical positivist types—define facts and laws as forms of statement. Molecular biologists even speak of a "central dogma." Consequently, facts are not things, pure and simple, separate from language, and while laws may describe natural regularities, formally they are less like rocks and more like rules, propositions, and maxims. Similarly, a scientific "discovery" is not the thing discovered. To call a thing a "discovery" implies a relation to an historically changing context of scientific activity. A discovery is something that had not been discovered before, at least not by scientists (Brannigan 1981). I have never seen an adder, and if I were to stumble upon one during a walk through a patch of heath, I might be startled, frightened, and amazed, but I would not be stupid enough to claim a *scientific* discovery. A discovery is a *temporal* and *communal* matter. Like a patented invention, a discovery is answerable to criteria of novelty, significance, and intelligibility, as judged by credentialed members of a disciplinary community. So, whenever scientists *say* that something is an objective fact or law, and whenever they claim that something is a discovery, they are engaging in communicative actions in particular organizational and historical contexts.

Philosopher Peter Winch (1958, 85) pointed out that when physicists describe their experiments and instruct others how to perform them, they employ a language of *action* for communicating results and instructing others on how to do experiments. Weinberg begins to acknowledge something like this when he says that "the languages in which we describe rocks or in which we state physical laws are certainly created socially," but as Winch makes clear it is not just that a language is *created* socially, it is used in day-to-day activity within a community. A scientist's words are not just a set of references corresponding to an object, they are written and spoken expressions for others to read, hear, and understand. In certain respects, following such instructions is like learning a game that involves specialized equipment, rules, skills, judgments, improvisations, and characteristic expressions. When we begin to think along these lines, the baseball analogy no longer seems so farfetched. To say that the sciences are like games need not imply that physics is *nothing but* a game like baseball. The challenge raised by the baseball argument is a matter of seeing how far the analogy can be pressed before it becomes absurd. And, like the airplane argument, the baseball argument verges upon some deeper questions: What is the relationship between language and the world? Are physical laws like rocks in the field, or are they more akin to rules governing human actions? To what extent are scientific methods like other games? These questions are related to larger historical questions about the origins of the European sciences and smaller questions about particular historical events. The participants in the more splenetic exchanges associated with the science wars often fail to acknowledge the long history of debate, the extensive background of scholarship, and the chronic uncertainty associated with these questions.

CONCLUSION

Like any metaphor, the science wars analogy can be further extended. It would be possible to imagine the invention of "smart bombs" aimed to eliminate true antiscientists while sparing more benign constructionists, or to identify casualties from "friendly fire" within each armed camp. But, as I have suggested, there is at least one large disanalogy associated with the science wars metaphor. The participants use fighting words, and some may have influenced editorial judgments and university recruitment and promotion decisions, but for the most part the "wars" have taken the form of metaphysical debates conducted by nonphilosophers. When discussing the airplane and baseball arguments, I connected them with a series of larger questions on the relationships between language and reality, science and common sense, and social rules and physical laws. In this essay, I have not tried to resolve such questions, nor have I suggested that it is possible to resolve them. When faced with such questions, those of us who are caught up in the science wars can choose among a number of options. Many of us are not credentialed philosophers. We are natural or social scientists who, to use another baseball analogy, play sandlot philosophy with pick-up teams. Consequently, we may decide that any resolution of the questions we have been debating should be left for the professionals. Perhaps we would want to consult professional philosophers to lead, advise, or tutor us. Or, we may decide to forget about such questions altogether and get on with our empirical research. Those of us in the science studies field would face a dilemma if we tried to do so, because our empirical research involves descriptions of historical controversies, ethnographies of contemporary scientific practices, analyses of the impact of intellectual property legislation on scientific research, and other topics that place us in the thick of questions about relations between science and common sense, judgments about discovered and constructed entities, and debates about the extent to which "politics" penetrates scientific research and argument. These can be lively matters for the scientists we study, but not necessarily at the level of abstraction with which philosophers contend with them.

There is a profound difference between, for example, an epistemological puzzle about the difference between natural and constructed entities and a particular question about whether a feature seen in an electron microscope is a natural property of a cell or an artifact of the preparation. A microscopist faced with the latter kind of question can perform a series of procedural checks, which will resolve the question for all practical purposes. The practical solution might not put a skeptical philosopher's doubts to rest, but this would have little bearing on the adequacy of the microscopist's judgment about the anatomical feature. As Sharrock and Anderson observe: "The epistemological skeptic, who denies that we can ever really know anything, has no interest in getting into a dispute with someone who, say, claims to know where to find a good Chinese restaurant in a strange town, over whether they can in fact find such a restaurant" (1991, 51). It certainly is important to know the philosophical background, but the natural and social scientists who are currently involved in debates with each other have some advantages over philosophers when it comes to examining particular historical cases and situations of practical judgment.

My peaceful recommendation is not to end or resolve the debates about science, which have been mischaracterized as "wars." Instead, it is to continue the debates, while taking advantage of the educational opportunity they offer. The science wars have attracted a great deal of public attention. The debates offer an opportunity for participants, their students, and members of larger audiences to consider and reconsider questions, arguments, and perspectives that they previously ignored or rejected without due consideration. At least some debates between physicists and sociologists (such as the one between David Mermin and Harry Collins and Trevor Pinch in *Physics Today* [1996; 1997]) seek mutual understanding and education. With such examples in mind, I will end with some modest proposals for upgrading the level of argument, and argumentative conduct, in this great debate about science:

1. Scientists and social constructionists alike should be more careful in distinguishing metaphysical arguments about "Science," "Truth," and "Reason" from historically specific cases of scientific research, particular truth claims, and specific reasons given for a judgment. Constructionists should (and sometimes do) recognize that skepticism about positivist, rationalist, and other philosophical conceptions of Truth, Reason, and Science has little direct bearing upon the adequacy of particular factual claims. Scientists who publicly defend Science and Reason should recognize that they are confronting an opposition that also draws many of its arguments from long-standing metaphysical disputes.

2. Sociologists, feminists, literary theorists, etc. who make substantive arguments about biological or physical phenomena should take pains to know what they are talking about, just as physicists and biologists who make philosophical, historical, and sociological claims should be held responsible for being well informed about the relevant scholarship.

3. Disputes between social historians and scientists about particular historical matters of fact should be recognized for what they are: disputes about the historical record to be subjected to the usual standards of historical evidence and argument.

4. All parties should be wary of using terms like "merely" and "nothing but" when characterizing their own and their opponents' claims and analogies.

5. All parties should be reluctant to discount the views of opponents by trading in notions of false consciousness, pathological denial, ideological blindness, sheer stupidity, intellectual dishonesty, and the like. (And, of course, we should all try not to be false, pathological, ideologically blind, stupid, and dishonest.) Such attributions may turn out to be well founded in some cases, but they should be resisted in the interest of peace.

Chapter 5

CAUGHT IN THE CROSSFIRE? THE PUBLIC'S ROLE IN THE SCIENCE WARS

Jane Gregory and Steve Miller

"The public" is a mysterious entity, even to those experts such as sociologists, psychologists, and anthropologists whose job it is to study it. It is perhaps therefore not surprising that scientists, who have, like anyone else, lay knowledge built from personal experience and common sense, tend to describe the public as a very different phenomenon from the one that continues to challenge the professionals. The irony of the situation is, however, that scientists' conception of the public consists of people who are opposed to science because they are either willfully ignorant of or vulnerably deficient in scientific knowledge, while the social scientists' researches describe a public that trusts in science, acquires and mobilizes relevant scientific knowledge as and when needed, knows a daft idea when it sees one, and generally holds science in very high regard. The question is, then, why would scientists—whose own work is a manifesto for the insufficiency of personal experience and common sense (the Sun, they tell us, does not go round the Earth despite appearances to the contrary)—cling to their own negative and worrisome perception of the public when the sociologists offer a much healthier picture of public attitudes to science?

One reason for this obstinate pessimism in the face of the positive findings of sociological research is that when most scientists think about public understanding of science, they do so within the framework of what has become known as the "deficit model." Within this model, scientists are the providers of all knowledge, and the arbiters of just what should be provided, to an empty-headed public. This approach has many attractions for scientists: it places them at the top of the epistemic heap; it leads to a clear program of action—"Fill those heads!"; and, the results of the program should be empirically testable in surveys of public understanding of science. But it is an approach that fails woefully when the science concerned is disputed within the scientific community or is in the process of being developed—when the science is science-in-the-making. And this is where the problem lies, because much of the science the public needs to know about is either hotly contested or is still on the assembly line, while scientists try to work out just which bit goes where. Given that this science-in-the-making is precisely the kind of science about which scientists are least sure and in which sociologists of science are most interested, and given that this is the most public of science, it is no wonder that the public have been dragged willy-nilly into the science wars.

ENLISTING THE CIVILIANS

The meaning of the phrase "public understanding of science" has evolved and diversified in the postwar era: what used to refer to the little-understood and barely interesting phenomenon of the conception of sciences among laypeople now serves a variety of purposes. It provides a label for normative and operational definitions of what the public understands about science, as well as for policy in the area and for the social and educational movement the idea has spawned; the term is also a job description and a sphere of research and practice for academics and communicators. It is also the name of a peer-reviewed scholarly journal, which began life in 1992 with an editorial board whose membership included physicist and historian of science Gerald Holton and sociologist Harry Collins; its first edition carried an article by Holton entitled "How to Think about the 'Anti-Science' Phenomenon" (1992), a paper which evolved into a chapter of Holton's *Science and Anti-Science* that was published the following year (1993, 145–89). Collins's and Trevor Pinch's *The Golem,* also published in 1993, is subtitled *What Everyone Should Know about Science.* These became important position statements on the public understanding of science on opposite sides of the science wars.

Alongside the increasing interest in the more programmatic aspects of this latest surge in public understanding of science activity developed a boom in popular science broadcasting and publishing, epitomized perhaps by the success of Stephen Hawking's *A Brief History of Time.* But this boom also brought a definition of science into the public domain that was broader than the scientific establishment had envisaged when they urged the media to carry more science. Thus science television in the UK, for example, now looks beyond sturdy documentaries about nature and technology to explore fields such as the social, cultural, and ethical aspects of science, as well as more contentious areas such as New Age science and the paranormal. Paul Kurtz, chair of SCICOP, the Scientific Committee for the Investigation of Claims of the Paranormal, notes—with alarm—similar developments in book publishing in the United States, where "if one visits any bookstore in America, one will find that the New Age, inspirational, spiritual, and paranormal shelves far outdistance the science books made available" (1996, 495).

The prominence of New Age science and parascience in the media is perceived by some as evidence of an "antiscience" movement. Central to their arguments is the notion of "rationality." Holton's original paper, for example, counted a resurgence of belief in astrology as symptomatic of a decline in our collective rationality; and evolutionary biologist Richard Dawkins, among others, has shaken his rationalist fist at what has become seen as the paradigm of paranormal propaganda, *The X-Files:* in a televised lecture in 1996, Dawkins—speaking in his role as Professor of Public Understanding of Science at the University of Oxford—felt this media preoccupation with "pseudoscience" threatened the proper appreciation of science and dampened the "appetite for wonder" that it could satisfy. Television channels, he warned, were unleashing an "epidemic of paranormal propaganda" that threatened to take us back to "a dark age of superstition and unreason, a world in which every time you lose your keys you suspect poltergeists, demons or alien abduction."

What Dawkins longs for is that we should all think like scientists, and he is far from alone in this. In 1995, the New York Academy of Sciences played host to a conference entitled "The Flight from Science and Reason" (Gross, Levitt, and Lewis 1996) held to rally American academics to the banner of "defending science" raised a year earlier by Paul Gross, the former director of the Woods Hole Marine Biology research station, and mathematician Norman Levitt. Levitt's contribution to the conference was an impassioned plea for mathematical thinking to become commonplace, for mathematics, after all, is

> the kind of mental calisthenics where principles of logical inference are rigidly applied, ambiguities of language sifted out, unstated assumptions kept from the premises, words prohibited from sliding unannounced from one meaning to another, and appeals to emotion or cultural prejudice or moral indignation despised. . . .

> The real value of even a modest mathematical education is that it breeds a certain salutary impatience, a distaste for intellectual flatulence, for otiose pseudotheorizing, for argument by browbeating. It breeds a certain shrewdness, as well, in all sorts of odd corners of modern life. It helps purge the staleness, the laziness, the careless propensity to accept unexamined cliches, from one's thinking. (Levitt 1996, 47)

This aim of reasoning like a scientist is to be found explicitly or implicitly in many strategies for public understanding of science. The idea is not new: in the 1930s the US educationalist John Dewey (1934) urged that children should develop a "scientific attitude" that would guide them rationally and logically through the trials of everyday life.

Systematic attempts to determine levels of what—following Dewey—came to be known in the United States as "scientific literacy" began in 1972, when the National Science Board started its biennial *Science Indicators*—social surveys that gauge people's knowledge and understanding of, and attitudes to, science. According to political scientist Jon D. Miller, who for many years administered and analyzed the surveys, the data for 1979, for example, indicated that only 14 percent of Americans could give a satisfactory answer when asked what it meant to study something scientifically—an answer that included the notion of testing, and modifying, hypotheses by, and in the light of, experiment (Miller 1987). This number was not significantly different from the 12 percent who could achieve this level of scientific literacy in the first such survey, administered in 1957. And, when those of 1979's 14 percent who "failed" other tests were discounted (such as those who believed that astrology was at least "sort of" scientific), the overall percentage of scientifically literate Americans halved to 7 percent. In 1985, the figure was 5 percent, and similar levels have been recorded in all subsequent surveys.

Social psychologists Martin Bauer and Ingrid Schoon (1993) have pointed out that surveys such as these, which aim to measure the extent to which laypeople think like scientists and know what scientists know, will inevitably categorize the vast majority of people as scientifically illiterate for the simple reason that most people are not scientists. Nevertheless, these surveys have been interpreted not as revealing this simple fact, but instead as exposing the horror of scientific ignorance—the void in people's minds where the science ought to be. Again the deficit model underlies the belief that these empty heads will readily absorb the antiscience peddled by fringe scientists, the mass media, and other forces of irrationality; no wonder, then, that the credulous public fall for crystal healing, pyramid power, and *X-Files* special agents Mulder and Scully.

However, the deficit model has been subject to much criticism. One of the earliest critiques was written in 1952 by the distinguished historian of science I. Bernard Cohen, who listed what he called the fallacies in arguments for improved public education in science. Among these was the fallacy of critical thinking, which, Cohen argued, is not necessarily inculcated by a scientific education, as "may easily be demonstrated by examining carefully the lives of scientists outside the laboratory." Those who despair at the poor results of knowledge surveys might take note of Cohen's fallacy of miscellaneous information: "the belief in the usefulness of unrelated information such as the boiling point of water . . . the distance in light years from the earth to various stars . . . the names of minerals." Cohen was also concerned about the fallacy of scientism and argued that science is not necessarily the only or even necessarily the best way to solve many sorts of problems. Despite his criticisms, Cohen's agenda was proscience: he looked forward to a day when journalists and educators would take up his challenge, for "that science can make the decisive contribution to the physical existence of man is without question" (Cohen 1952).

In addition to these criticisms, the deficit model finds little empirical support from recent fieldwork by researchers in social psychology, anthropology, and sociology. They conclude that rather than toting empty heads, laypeople have minds that are chock-full of intellectual strategies for dealing with the problems of everyday life, of which scientific strategies may not be the most important. Laypeople use different mental languages from scientists and work with what Susanna Hornig (1993) has called "an expanded vocabulary": far from operating with limited or deficient machinery, laypeople mobilize a broad array of tools to solve problems through science, culture, emotion, ethics, morality, trust relationships, and customs. These may be small tools—a Swiss army knife of small instruments compared to science's mighty saber—but they cut through the tangle of contemporary existence nevertheless and produce solutions that sit more easily with people's lives and consciences. This broad contextual approach to problem-solving gives its name to the most powerful alternative to the deficit model; the contextual model is one that treats the public as experts to varying degrees and in a variety of ways in their own sphere—a sphere in which the *scientific* experts would tread much more constructively, it is argued, if they could think not only like scientists but also like the public.

SHOOTING THE MESSENGERS

The media play a prominent role in stories of the public in the science wars, and because media contents are more accessible than the public mind, the two have become conflated in ways that serious media researchers would not contemplate. For example, the proliferation (so it is claimed on the basis of anecdotal evidence) of media accounts of astrology and the paranormal is taken as sufficient evidence that the public is correspondingly interested in these phenomena, or even persuaded of their real value. In his introduction to *The Flight from Science and Reason,* Paul Gross bewails the "widespread promotion" of "humbug and credulousness" in "the respected media of communications" (1996, 2). SCICOP's Paul Kurtz claims that "the mass media continually feeds a receptive public with modern tales of paranormal miracles," and that "the mass media are in the business of packaging and selling the paranormal as a product" (1996, 495). These claims reflect a poor understanding of the relationship between science, the media, and the public.

First, the mass media representation of science is overwhelmingly positive. What the public sees is, according to sociologist Leon E. Trachtman, "an essentially positivist portrayal of science as a heroic, apolitical, and inherently rational endeavor" (1981, 12). Science communication scholar Dorothy Nelkin (1995) and others have argued that the close allegiance between science journalists and the scientific community has produced science coverage that is more deferential to its subject matter and constituency than would be acceptable in other fields of journalism. This has resulted in the representation of science that Trachtman describes—one that ignores both the contingency of scientific knowledge and its social and political context. Although most scientists are happy with this image, it has its dangers. As David Goodstein pointed out in his contribution to "The Flight from Science and Reason" conference: "I think we scientists are guilty of promoting, or at least tolerating, a false public image of ourselves that may be flattering but that, in the long run, leads to real difficulties when the public finds out that our behavior does not match that image. I like to call it the Myth of the Noble Scientist. . . . [T]he ideal scientist [is] more honest than ordinary mortals, certainly immune to such common failings as pride or personal ambition. When it turns out, as it invariably does, that scientists are not like that at all, the public we may have misled may react with understandable anger or disappointment" (1996, 37).

Second, implicit in claims of public interest based on media activity is an assumption of the media's power to create beliefs and attitudes in a credulous public. This is not a claim most effects researchers would feel comfortable with; the ongoing debates about the influence on viewers of media violence and the impact on voters of media politics indicate that such research has some way to go before it can provide us with any conclusive insights into the complicated interactions between people and the media (Gauntlett 1995). In the case of the public and the paranormal, assumptions of a media effect relieve us from the cumbersome necessity of researching public attitudes to the paranormal: we can simply use the amount of media material as a gauge. However, this assumption of a media effect can also demolish our gauge, because if the public is persuaded by powerful relentless media renditions of parascience, isn't the public just as vulnerable to the even more substantial quantities of real science in the media? Even the British tabloids contain five or so science articles for every astrology column; if it's the medium that's powerful, those rational science articles are persuasive too.

Thus the simple presence of information or attitudes in media products is a very poor indicator of public knowledge or opinion. So when, in the science wars, Kurtz reports that "virtually every major newspaper carries an astrology column" (Kurtz 1996, 495), his observation cannot be faulted; but what his observation cannot tell him is how people react to these columns. One recent study, for example, suggested that while people with very high levels of scientific knowledge tend not to believe in astrology, the correlation between knowledge of science and belief in astrology is much less clear cut at lower levels of knowledge—there, knowing a bit more about science doesn't necessarily make someone more skeptical of astrology. The study also found that while 73 percent of a representative sample of two thousand British adults said they read their horoscope, only 6 percent said they took astrology seriously (Bauer and Durant 1997). It is shocking indeed if a US president resorts to astrology for political advice; it is comforting perhaps that mostly the public aren't that stupid.

Another relevant question is the extent to which the mass media create, rather than reflect, popular culture. Certainly the popular interest in the paranormal is not a media creation: ghost stories and fortune-telling have long been a feature of Anglo-Saxon culture, among many others. Horoscopes first appeared in UK newspapers in 1930 when the *Sunday Express* used them as the basis for a story about the newborn Princess Margaret; astrology flourished during the 1930s when political and social unrest made people concerned for their future, but it was not a time of antiscience in the UK, as evidenced by the success of scientist-popularizers such as J. B. S. Haldane, Lancelot Hogben, and J. D. Bernal.

Similarly, the proliferation of astrology columns in newspapers in postwar Britain owed more to the repeal of the Witchcraft Act in 1951 than it did to any shift in public opinion against science—science was also widely celebrated in the postwar years. In any case, this interest in astrology was neither new nor generated by the newspapers: the public has long been visiting astrologers in fairground booths and suburban parlors, and the only difference now is that no one has to be furtive about it. Writing toward the end of the Glasnost period in the USSR, physicist Sergei Kapitza reported (1991) that the social upheavals engendered by the breakdown of "scientific socialism" had caused a sharp turn toward extreme mysticism and the paranormal. But one can wonder to what extent this public interest was revealed, rather than created, by the demise of the Soviet regime.

COUNTING THE DEAD . . .

Our remarks so far concern public understanding of science as a field of inquiry, rather than as the phenomenon in laypeople's heads. So: did the science wars as an episode make any difference to the ways in which the public understands science in the same way as episodes such as BSE or NASA's "Martian" meteorite impinged on public understanding? Our suspicion is—we have not researched this phenomenon in any systematic way—that in this sense the science wars made not one iota of difference to the public understanding of science. The public does not read *Social Text;* only a small fraction of the public even reads the *New York Times.* In the UK, the science wars arrived over the Internet, but even then only to Internet users who knew where to look. Alan Sokal may have beamed at British citizens from the pages of a *Guardian* supplement, but no one really noticed.

However, this is not to say that the science wars failed as a public relations exercise, for its public was not the general public but specific publics which, under many definitions of that tricky term, would not count as "public" at all. When Levitt calls for a broader appreciation of the delights and strengths of mathematics, he is explicitly calling for this appreciation among "one's fellow intellectuals" and "the broader intellectual community"—it is those across campus, not those outside of it, who are the focus of concern here. And when professor of medicine Wallace Sampson (1996) worries about alternative healers, he is concerned more for the medical orthodoxy they reject than about what they might be doing to their patients. There are those, such as Holton, whose sympathies, experience, and insights extend beyond disciplinary boundaries, who offer a deeper and more complex picture of science in public, which is, we suspect, closer to the truth; but this picture is, because of its depth and complexity, less useful as a manifesto in the kind of battles conducted during the science wars. It is a matter of regret that the temper of the debate made subtlety the first casualty of the war.

But even someone with Holton's social sensibilities can underestimate the public and overestimate the impact of particular episodes in shaping the public understanding of science. For example, the considerable concern expressed by Holton and others that the Smithsonian Institution's 1994 exhibition *Science in American Life* was antiscientific and would engender antiscientific sentiments in its visitors was misplaced. The dispute over *Science in American Life* was precisely one of those occasions when the science war battlefields encompassed the public domain, and they did so because the science was displayed in its social and historical context, that is, in precisely the way the public experiences science. For example, one exhibit showed chemists arguing over how to market a discovery, and another, how American families coped with the threat of atomic war. Physicist Robert Park objected to this contextual representation of science and wanted a rather different exhibition: "[I]t is science that uncovers the problems, and it is to science that we turn to put it right. Not because individual scientists have any claim to greater intellect or virtue, but because the scientific method sorts out the truth from ideology, fraud, or mere foolishness. What people need to know, and are not told, is that we live in a rational universe governed by physical laws. It is possible to discover those laws and use them for the benefit of humankind" (1994, 207). Park feared the visiting public would go away with the message "that Western civilization is heavily burdened with guilt, and science, as a servant of a power structure, must bear a large share of that guilt" (207).

Museum director Alan Friedman visited the exhibition and noted, on the other hand, that public concerns about science were represented alongside the achievements of scientists; the exhibition was about science *in American life* and as such would necessarily embrace the reservations as well as the celebrations: "I find it hard to dispute that the public's attitude towards science in American life is far more cautious today than the earlier, near-Utopian view of a science-based society portrayed at the pavilions of the 1939 World's Fair. . . . Whether or not more positive case studies had been included [in *Science in American Life*], however, the exhibition would still have the burden of interpreting how American public opinion developed into today's respect for the achievements of science combined with skepticism for everything offered as a 'technological fix' to a social need" (1995, 306) Friedman also noted that "*Science in American Life* does make a substantial effort to portray scientists as caring human beings, striving to be good citizens and to help their compatriots in time of need" (306). Science communication scholar Bruce Lewenstein has reported (1996) that not only did the Smithsonian's own visitor study show that attitudes to science were "overwhelmingly positive" both before and after visiting the exhibition, but also that many commentators who were themselves critical of what was on display acknowledged that the average visitor did not interpret the exhibition as being negative.

The public, then, are not quick to think the worst of science and scientists, despite the fears of some. While the deficit model of public understanding of science offers up empty-headed, gullible people who might be expected to rally to which ever cause—scientific or antiscientific—is being proclaimed the more loudly, the contextual model reminds us that empty heads are very few and far between and that the public is constructing its image of, attitudes toward, and knowledge about science throughout long lives lived in the natural and technological worlds in which schoolteachers, grandparents, dentists, library books, commercials, food packaging, gadget manuals, friends, accidents, weather, hospitals, and so on are all contributing to a public understanding of science which, surveys repeatedly show, is overwhelmingly positive. Viewers in the UK have recently enjoyed the one-hundredth episode of *The X-Files,* but even its total broadcast output will amount to the tiniest drop in the enormous, complex ocean of influences in the public understanding of science.

. . . AND PROVIDING FOR THE SURVIVORS

Elsewhere we have argued that one can draw practical lessons from the public aspects of the science wars, and from social and historical studies of public science, of relevance to the public understanding of science, and we have tried to draw these together into a "Protocol for Science Communication for the Public Understanding of Science" (Gregory and Miller 1998, 242–50). Among its recommendations, this protocol suggests that science communication should be a process of negotiation between the scientific community and the public. This means that scientists must abandon the notion that all they have to do is to fill up empty heads. Instead, they should approach communication situations as a mutual getting-to-know process, a dynamic exchange in which disparate groups find ways of sharing a single message. This exchange has implications for the public, too, in discussing and deciding what they need from science: the public cannot expect their needs to be met if they do not articulate what these needs are.

We also argue that the key to the relationship between science and the public is trust, and that trust is established through the negotiation of a mutual understanding, rather than through statements of authority or of facts. Among other things, that means that while science has every right to defend its role as a provider of "reliable knowledge" in our society, scientists need to make clear that one of the key features of science is its inherent provisionality. And, in the area of science-in-the-making, this provisionality is the *essential* feature of scientific knowledge. In our view, scientific information has to be transmitted in ways that take proper account of the various social factors which make up the context of its transmission: these include those (often denied) social factors which go into its construction, as well as the knowledge and beliefs that the intended recipients already have, or have access to. Only this kind of science communication can facilitate public participation in the debates about the scientific issues that matter.

The business of science has, of necessity, become highly professionalized. But at the same time, the results of science have been more and more socialized in terms of the impact they have on the public, through changing beliefs, practices, or lifestyles. Scientists thus have a duty to explain their work to the best of their ability and to be open about the potential, limitations, and practices of science. They also have a duty as citizens to warn and argue privately and publicly against the misuse of their work. They are, surely, among the better placed to do this.

In the end, however, questions of which research will be supported and how the results will be used should—and will—be decided in the public sphere. That places considerable responsibilities not only on the scientific community and professional communicators, but on the public as well. If citizens are to fulfill the responsibilities in this area, communications that they receive about science—at least when that science is provisional and potentially controversial—have to be designed to facilitate their participation in the social processes of debate and decision-making. The type of communication designed to bring about an awe-struck admiration for the mysterious men in white coats is not what we need for the challenges of the twenty-first century.

THE TERMS OF THE PEACE

At the beginning of this chapter we pointed out that surveys of public knowledge of and attitudes to science consistently show that, while levels of knowledge may not be as high as scientists would like, attitudes could hardly be more positive. It would therefore appear that ignorance does not necessarily lead to antiscientific attitudes. Nor, it seems, do sociologically informed representations of science, like those in *Science in American Life,* nor media pseudoscience such as *The X-Files,* undermine the public's appreciation of scientists; if they do, there are plenty of more conventional representations to balance the picture. By pointing up the context of scientific discovery and highlighting the human aspects of the enterprise, the more sociological interpretations of science may reach people more readily and strengthen the place of science in the broader culture.

The enormous social impact of science has more than earned it a prominent place in the public sphere, and if scientists are serious about keeping it there, then they would do well to undertake the important social enterprise of public understanding of science hand in hand with those who can offer them some professional help. The public sphere is a sociological entity, and its workings are most transparent to sociologists; communicating with the public is something the media do very effectively. Surely, then, public understanding of science offers an area where scientists, sociologists, and the media could become allies. Individually, these groups have their own agendas, of course, and like every relationship, this unlikely threesome will thrive only with a great deal of negotiation and compromise. Its enormous potential, however, ought to make it worth the effort.

Chapter 6

LIFE INSIDE A CASE STUDY

Peter R. Saulson

It's been just my luck to live my entire scientific career inside a case study.

I'm a gravitational wave physicist. Since I finished graduate school in 1981, I've been engaged in the search for gravitational waves, an elusive phenomenon in need of verification ever since Einstein proposed the idea around 1916. When I joined the field, dozens of pioneers were already engaged in the quest. Now, we number many hundreds. That growth has come about because it may be soon that the quest succeeds, and the nearness of success breeds excitement. It is also the case that the method most likely to lead to successful detection of gravitational waves, laser interferometry, is pretty complex and needs a large team if it is to succeed.

My colleagues and I are drawn to this field because it promises several exciting scientific achievements. First, it will help cement our understanding of gravity by completing the demonstration of a key idea from Einstein's general theory of relativity. Second, success in detecting gravitational waves will automatically start providing us with new kinds of information about the distant and exotic astronomical objects that generate the waves, such as neutron stars, supernovae, black holes, and perhaps the Big Bang itself. In other words, we hope to move on from doing interesting physics to doing very interesting astronomy.

But it has to be said that the pride and excitement that we feel about our work can't avoid being mixed from time to time with something like embarrassment at the origins of our search. The active search for gravitational waves was begun by Joseph Weber in the early 1960s. By the end of the decade, he was claiming to have succeeded by having registered coincident minuscule excitations of a pair of widely separated aluminum cylinders that were each isolated from every other kind of disturbance.

Weber's announcement was greeted with a mixture of contradictory attitudes on the part of various scientists. Some disbelieved them because it would take stupendous energy releases to generate signals of the strength Weber said he saw, enough to exhaust the entire stock of matter in our Galaxy during the amount of time it has been around. Others were thrilled that such dramatic events nevertheless actually appeared to be happening. What united most scientists was a feeling that this was too big to leave alone; Weber's claims had to be followed up, either to refute them or to follow the path to dramatic new knowledge to which they appeared to lead.

By the mid-1970s, several rather close copies of Weber's apparatus had been built and operated. None of them detected any gravitational waves. This led to an acrimonious set of exchanges with Weber while an explanation for the contradictory results was sought. In the end, almost all of the participants concluded that there were no gravitational waves arriving at detectable levels, in spite of Weber's claims to the contrary. Weber was not convinced and, until his death in September 2000, continued to record data that he said vindicates his claim.

Some of the participants in this episode became disgusted or discouraged and left the field for other pursuits. Others became convinced that they could do much better at detecting the weaker gravitational waves that were actually expected. These latter became the nucleus of the field that I joined a few years later.

Such a dramatic instance of scientific claim, attempted replication, and final resolution of the claim has all of the elements of a perfect test case in the practice of science. Sure enough, it attracted the careful attention of sociologist of science Harry Collins, who carried out an intensive series of interviews with most of the participants in the affair, while it was being played out. The lessons he drew formed the basis for a number of his publications (1975, 1981d, 1992).

I became aware of Collins's work many years after his initial involvement in the affair. A visitor told me he was surprised to hear that my research was on gravitational waves, because Collins's 1985 book *Changing Order* (Collins 1992) had left him with the impression that the field was dead. This turned out to be a serious misreading of the work, but I was grateful to be enticed to read it nevertheless. The capture on audiotape of the rich variety of opinions held by practitioners in the midst of this confusing and dramatic episode is a great service to the understanding of science.

A conservative student of science might see strong confirmation of the standard model of the scientific method in the way the Weber affair played out. Attempts at replication failed, thus causing the abandonment of the original claim. Collins put a more provocative interpretation on the story. He noted that for several years, while negative results started to pile up alongside Weber's supposed detections, it was unclear whose experiments should be considered faulty. Indeed, how does science proceed in a situation where Experiment appears to speak with more than one voice? Collins declared this a common problem in frontier research and dubbed it the "Experimenter's Regress." In this interpretation, it is impossible to know by objective criteria alone whether one or another experiment has been performed competently. Thus, rather than providing an unambiguous way out of controversial affairs, experiments can only serve to reinforce the apparent conflict. Instead, says Collins, some external (i.e., social) event must occur to crystallize a consensus interpretation. Indeed, in the Weber affair it is easy to point to the polemical work of one of the scientists involved as the event that caused confused opinion to crystallize around the idea that it was Weber's experiments that were faulty.

It shouldn't be too surprising that a position that appears to denigrate experiments as a source of objective guidance in science has won Collins few friends among scientists. I actually like it, though, since it emphasizes what many experimenters know but don't like to share: that experiments are in fact hard to do well, and that it is fiendishly difficult to know when you've done one correctly and can stop looking for errors. But I found other ideas to contest in Collins's theory of how science proceeds, in particular his silence concerning the nature of the intellectual debates that accompany the social process of crystallization of a scientific conclusion.

So I wrote him a letter. This is something I almost never do. But I was intrigued by his notebooks full of interviews with my senior colleagues. I also thought that someone who had recognized the anxieties of an experimenter might be educable where he disagreed with me on the functioning of the scientific process.

Collins wrote back and pointed out that he was soon to visit his colleague Trevor Pinch in Ithaca, New York, not far from my home in Syracuse. There we struck up a friendship and also a professional relationship that has continued to the present. For as I learned, Collins had again picked up the story of gravitational wave detection efforts, both to follow the survivors of the Weber affair and to trace the ascendancy of interferometers, which may soon close the story of controversy and open a new volume of normal science. Since that first meeting, we have talked many times at scientific conferences and meetings of the LIGO Project (the experiment with which I'm associated). Our relationship is a mix of the asymmetrical one between sociologist and informant and a more symmetrical one in which we debate the nature of the scientific enterprise.

Collins describes himself as a scientist whose object of study is the social process of science. This claim exposes a sympathy for scientists which too casual readers of his work might be tempted to suspect (unfairly) he lacks. It also puts him in the uncomfortable position of being a scientist whose subjects talk back to him. As a physical scientist, I'm happy not to have to worry that my professional actions might affect the well-being of the systems I study. I'm also quite glad not to be told by my subjects that I'm going about my work all wrong—colleagues can be relied on for that often enough to keep my work honest. Collins has to deal with both sorts of problems, with regard to his subjects and with criticisms from colleagues as well.

As an example of the latter, his interpretation of the Weber affair has been challenged by Allan Franklin, a former physicist who defends a conservative interpretation of the history of science. Franklin wrote a critique of Collins's interpretation (and methodology) of the history of early gravitational wave detection efforts (Franklin 1994). Franklin claims that the story is a classic case of experimental evidence deciding a scientific question. In a rebuttal published as a companion piece, Collins (1994) insists that a more subtle interpretation is required. He certainly would not deny that scientists could be justified in interpreting the evidence the way they have done. But, he claims, that interpretation is not *required* by the evidence alone. A simple algorithmic process called "scientific method" does not force a single interpretation of experimental claims. Rather, experiments suggest several possible interpretations, thereby underdetermining the outcome of controversies. It falls to the mysterious social process of consensus building to find the interpretation that is eventually held to be true.

The issue is not one of merely academic interest. All of us present actors in gravitational wave detection have had to form our own interpretation of the history of our field. And on that interpretation stands our notion of the value of our work, from the validity of the scientific choices we make from day to day to the reasonableness of the case we make that our field of research is worthy of the not inconsiderable sums that the taxpayers of this and other countries have contributed to support our work. It is an often-expressed feeling that we need to be extra careful as scientists in order to redeem our field from the stigma of having been created out of unsubstantiated claims. Others feel the opposite, that too great a worry about the supposed sins of the past creates too much caution where exploration instead would reward boldness over care. And memories are still vivid; most of the players from the seventies are still alive, and quite a few are still active in the field. But it must be said that the level of this debate is more like a steady simmer than a rolling boil; from day to day it is indeed possible to ignore it.

Collins's renewed presence among active scientists is yet to be fully digested by the community. So far, he has tended to renew contacts with the veterans of the earlier days, while gradually including younger people in his circle of informants. The veterans came to trust Collins for his objectivity and his commitment to shielding the identities of his informants, so even while they may not care much for his interpretations (or in many cases care anything about them), they are comfortable with his presence.

The many outside of this circle mostly have taken a cold-eyed view of the possible difficulties that could ensue from sharing too much with someone who might not share our views or our values. In part, this is simply a principled suspicion of someone who challenges the value of the thing an experimental physicist holds dearest: experiments themselves. In part, it is resentment (expressed explicitly to me by one colleague) of being treated like a laboratory rat, while living in a system that as it has grown larger and more successful has reduced the human autonomy of many of its participants. Thoughtful scientists must also worry about any possible source of embarrassing revelations, given that we work on a project whose scientific risks and substantial costs are as well known outside our field as they are inside.

Thus I expect that some of the most interesting aspects of the relationship between Collins and his subjects are yet to play themselves out. Personal suspicion will almost certainly fade away as he becomes better known. But he can probably look forward to many vigorous debates one-on-one about the nature of science as he adds to his list of active informants.

But even while Collins has yet to fully carry out his project of making my colleagues and me the subjects of his work, his own previous work has made him a subject of public discussion of substantial intensity. In the pages of *Physics Today,* the monthly magazine of the American Physical Society, David Mermin of Cornell University has devoted several of his "Reference Frame" columns (1996a, 1996b) to a critique of *The Golem,* a book by Collins and his Cornell colleague, Trevor Pinch (1993).

The Golem is a work aimed at explaining to a popular audience the basic ideas of the work on the sociology of scientific knowledge carried out by Collins and Pinch. The basic theme is that science is not an automatic process that infallibly produces truth. Rather, it is a very human social process that, through the skillful actions of its practitioners, tends to garner useful knowledge better than any other process that we know. Collins and Pinch discuss a number of case studies drawn from their work to show how in each case the progress of science came through a more subtle process than a simply mechanistic use of experiments to adjudicate between competing hypotheses. There is a chapter on the early days of gravitational wave detection, drawn from Collins's previous work. But the case study that drew detailed criticism from Mermin was a discussion of the history of the establishment of the special theory of relativity.

It needs to be noted that Mermin's judgment on *The Golem* was far from wholly negative. But in his first column on the subject, he did find much to dispute in its chapter on relativity. The central issue was what follows from a demonstration that relativity was not simply established by one or two "crucial experiments" and, in particular, the meaning of the fact that Einstein was able to dismiss with a quip ("The Lord is subtle but He is not malicious") the claim made by D. C. Miller in 1921 to have measured a small ether drift effect in a version of the Michelson-Morley experiment. Mermin says that Collins and Pinch leave the reader with the impression that physics does not get its crucial guidance from experiments. What they should have said instead, says Mermin, is that experimental support for ideas in science comes not from one or two crucial experiments but from a "tapestry" of experiments, theories, and interpretations. The resulting interconnected structure of inferences is so resilient that seldom is the weakness of a single link relevant to the outcome of a scientific investigation.

A letter of reply from Collins and Pinch was published, followed by more discussion in Mermin's column, appearing to be a bit heated at times even as a rapprochement was being negotiated (Mermin 1996c, 1997; Collins and Pinch 1996, 1997). My own reading was that this was a battle of straw men. The idea of crucial experiments was an easy target for Collins and Pinch. In turn their demonstration of that paradigm's failure is clearly an insufficiently deep point to which to bring a lay reader in search of understanding of the process of science. This was the point of Mermin's discussion of the tapestry.

But was there a deep disagreement? I don't think so. *Changing Order* (Collins 1992) contains a whole chapter called "The Scientist in the Network" that explains how ideas become accepted through a process of negotiation between people defending various parts of the set of interconnected ideas that together make up our knowledge of the world. How different is this from the idea Mermin called the "tapestry"?

The dispute, to the extent that there is one, appears to lie chiefly over whether the language used to describe this process emphasizes the attempt to fit together many partial pieces of knowledge, or whether the language emphasizes that people with different interests are trying to do the fitting. Physical scientists on the whole are more comfortable talking about actions happening in the world of ideas, while social scientists prefer to describe the action as social.

Different styles of language use appear to be at the root of many of the disputes that have been loosely grouped together under the heading of the science wars. This was brought home to me as a consequence of having helped to organize an event at Syracuse University at which Alan Sokal and a panel consisting mainly of humanists discussed his hoax article in *Social Text*. For the better part of an afternoon, the panelists made a variety of attempts to put the affair in perspective. While each presentation was well prepared, I had the sad feeling that we were all talking past one another.

The key moment of the afternoon came when we invited questions from the floor. A graduate student of English stood up and, with obvious anger, asked Sokal, "What is your theory of language?" As she sat down, Sokal fumbled for an answer, clearly uncertain what the question meant. The matter was allowed to drop, and the event wound down to its conclusion.

It was only many weeks later that I realized what the question had meant, and why it was precisely the right question to ask. The key rhetorical device in Sokal's hoax article is to analyze for its truth value a single sentence, taken out of context, written by a famous person whom Sokal wishes to ridicule. Now if he were critiquing a piece of writing in the natural sciences, this might be a perfectly reasonable way of assessing the validity of the work. Scientific writing often has the character of a mathematical proof; if a key sentence can be shown to be false, then the whole argument will fail. This kind of language use is much less common outside the sciences (with the exception of philosophy). Thus, Sokal's technique was seen, by those used to a different kind of reading, to flagrantly miss the point of the ridiculed works. But neither he nor most of the other scientists in attendance (including myself) even heard the argument that was being made by the questioner.

Rhetorical tricks aside, there was a second question of language use at the heart of the Sokal affair. What offends Sokal and many other scientists as much as anything else is a perceived misappropriation of scientific language by people who don't "deserve" to use it. Along with the use of this kind of language is perceived to come a kind of authority separate from what is earned by the internal logic of the argument that is being dressed in this way.

Compared with the bitterness generated by the Sokal affair, the discussion over language that Mermin provoked with Collins and Pinch was both mild in tone and constructive in spirit. And in this case at least, it might be claimed that the social scientist ("nonscientist") used the more accurate and scientific language. A tapestry of facts and ideas is a beautiful metaphor, but a negotiation between people is a process that can be observed and described objectively.

Most of my colleagues are uncomfortable with any description of the scientific process that does not always keep in the foreground the particular scientific ideas that are at issue. We tend to be deeply suspicious both of the idea that our "interests" might be thought to influence our judgments as scientists and of the motives of someone who might make that claim. We tend to suspect a hidden (or perhaps even explicit) agenda, especially a political one.

One small incident in Collins's exploration of the community of gravitational wave experiments can serve to illustrate this point. At one point in our private discussions, Collins asked me if I had read any of several recent papers by Weber. I hadn't and cheerily admitted as much. Collins explained to me that this was completely normal, and we had an interesting discussion about the less formal ways by which scientific communication takes place. A colleague of mine had a different take on Collins's interest in the subject. This colleague had taken offense at a question on journal reading habits on a questionnaire that Collins had distributed at a scientific meeting. "What is he trying to prove," he asked me later, "that we just close our eyes to the evidence and make up what we want to believe?"

This defensive attitude can be seen in part as resentment at having to share with outsiders the job of explaining science to the lay public. Scientists have had a substantial monopoly on this job of interpretation for many years. Losing control of how we are presented to the public is uncomfortable. But there has also been just enough politically inspired misunderstanding of science to justify a feeling that scientists are surrounded by jealous and ignorant enemies, who will do us harm if we let them. Calls for "democratization" of science in favor of various well-meaning social projects have become a fixture of the discussion of science in some circles. (See some of the articles in the special issue of *Social Text* entitled "The Science Wars" [reprinted in Ross 1996a], the one in which Alan Sokal's spoof was published.)

To lump Collins with this latter movement is a disservice. Physical scientists need to learn to make a more nuanced reading of sociological writing. The "interests" that Collins ascribes to scientists in the network as they "negotiate" the acceptance of a new idea are for the most part a human embodiment of exactly the same process that Mermin would describe as the actions of individual threads in the "tapestry" of science. Except that science really is done by people, not by threads. So by naming them as people, we can investigate their behavior directly and not just by analogy. And on the occasions when individual scientists do act on interests that aren't purely intellectual (most scientists know of such occasions), that can be fit into a richer understanding of the socially embedded process that we call science.

At the same time, the work of Collins and others would be enriched if the discussions of "negotiations" among "interested" parties when interpretations of experiments are "crystallized" were clarified so that the primarily *intellectual* nature of the interests involved were made plainer. This would go a long way toward allowing all concerned to see concepts like the Experimenter's Regress in the proper light.

In the end, there may always be some unavoidable tensions between our communities, based on where our different curiosities lie. Physical scientists tend mostly to be curious about the subject of their scientific work, much less so about the social structure of the world in which they carry out their scientific work (except in the form of gossip, which is universal). To the extent that we are curious about how science works, it tends to be related to trying to understand reasons for the remarkable success science has had over the past few centuries in building an ever deeper account of the natural world. In search of explanations for this success, we tend to look for a simple overarching principle, such as the formula that science succeeds by looking for the simplest possible explanation. I have myself urged such formulas on Collins, without much success, but I am hardly alone.

Collins, on the other hand, is not nearly so much interested in the progress of science over the long run. His thrills come from the study of how scientists behave, as people, in the day-to-day work on the frontier where the answers are in doubt. There is certainly much to be learned here, and I feel I have learned some interesting things by paying attention to Collins's work.

A feeling of frustration among my colleagues comes from a sense that Collins loses interest in our work at exactly the point when we get it right [15]. We know that we grope as fallible people from day to day. But the dignity of our struggles comes from our knowledge that, time and again, a process just like the one in which we are engaged has led to progress. Seldom in the history of modern science has the community as a whole been misled for more than brief periods of time.

Collins, as a sociologist, is much more interested in scientific controversies than in the march of scientific progress. It is easy enough to see why this might be so. When evidence is clear and overwhelming, there is no longer an Experimenter's Regress. The behavior of all actors is then driven by this clear evidence, and there are no more interesting disputes or tensions.

But there is a sociological point of view from which precisely this boring state of affairs could be considered the most interesting question. Collins describes the network of interested scientists as incorporating in scientific discussions not only "internal" questions of science itself but also "external" interests of society. The metaquestion of the interest of society in maintaining a vigorous community of scientists (and supporting it, at substantial expense) is surely worthy of study. Is it not a powerful social imperative that in the end has influenced the maintenance of a social structure of science that can continue to successfully produce new knowledge? Would not a scientific system that too crassly bowed to perceived social demands, trimming scientific knowledge to the whims of politics, in the end do a disservice to the larger society? The feedback loop I am describing here is obviously a drastic oversimplification of a very rich set of mutual influences between science and the larger society. Nevertheless, here are a set of sociological or historical questions that would engage the actions in which scientists themselves take the most pride.

Chapter 7

CONVERSING SERIOUSLY WITH SOCIOLOGISTS

N. David Mermin

THE RELATIVISTIC GOLEM REVISITED

Several years ago I published a couple of columns (1996a, 1996b) in *Physics Today* criticizing how Harry Collins and Trevor Pinch treated the special theory of relativity in *The Golem* (1993), their book of case studies in the sociology of scientific knowledge that aims to tell the general reader "what you should know about science." The columns were followed by two exchanges in the Letters column (Collins and Pinch 1996; Mermin 1996c; Collins and Pinch 1997; Mermin 1997). I also had many conversations with both authors, by e-mail and face-to-face.[1] Recently Collins and Pinch have reflected on these exchanges in a lengthy afterword to the second edition of *The Golem* (1998a).

Reviewing all this, I'm struck by how much less we disagree than we appeared to at the time. Putting it another way, I'm surprised at how much we each originally misunderstood what the other was getting at. For example I now regard as a relatively minor matter what once struck me as the central question: whether the construction of scientific knowledge should be viewed as a process of discovering how nature works or as a process of consensus building among scientists. I'm increasingly persuaded that any issue one can formulate in one language has a paral-

Work on my contributions to this volume was supported by the National Science Foundation, grant no. PHY9722065.

1. Pinch is a Cornell colleague whose office happens to be in the same building as mine, and I have had extended conversations with Collins in Ithaca, Santa Cruz, Southampton, and Bath.

lel formulation in the other. The two points of view can lead to sharp disagreement on fundamentally metaphysical questions—for example, "Are electrons real?" And they can lead to strikingly different kinds of rhetoric about the nature of science. But the only substantive consequences of these different perspectives are the choice of issues they induce their adherents to investigate and which aspects of the scientific process receive the greater emphasis. Such differences are important, but they don't constitute grounds for anybody to doubt the good faith or common sense of anybody else.

I shall reexamine our exchanges on relativity, with a view to identifying avoidable ways in which we misunderstood each other, seeing what residual disagreements remain, and noting the many similarities with a more recent exchange I had with Barry Barnes and David Bloor (Mermin 1998a; Barnes 1998; Bloor 1998; Mermin 1998b) about their textbook *Scientific Knowledge* (Barnes, Bloor, and Henry 1996).

Unbiased but Damaging and Distorted?

My misunderstanding of Collins and Pinch began with the title of the chapter I criticized: "Two Experiments That 'Proved' Relativity."[2] I took the sneer quotes around "proved" as saying: We're going to tell you about two experiments that are said to have proved the theory of relativity, but in fact there is no proof. It became clear in our subsequent discussions, however, that the intent of Collins and Pinch was not to cast doubt on the theory of relativity, but only on whether those two experiments, the Michelson-Morley experiment and the Eddington solar eclipse expedition, provided the compelling support for relativity attributed to them in the folklore of physics. They used the quotation marks to indicate reservations about the definitive character of the experiments—not to indicate skepticism about relativity itself.

Primed, however, by my reading of their title to expect an argument that relativity rested on two experiments of questionable validity, I found it easy to read much of what followed as supporting this interpretation. In the second paragraph of the chapter, for example, readers are told some of the surprising elementary facts about relativistic physics, variously introduced by "If Einstein's ideas are correct" and "if the theory is correct." For me these qualifications indicated that an open-mindedness toward the validity of the theory was being urged on the

2. Most physicists would use a word like "confirmed" in this context, reserving the stronger term "proved" for a sequence of purely mathematical steps in a physical argument. But the ambiguity remains if "proved" is replaced with "confirmed."

reader. In Collins and Pinch 1996 they explain that in their own minds their narrative at that stage was already situated in an earlier time when the truth of relativity was far from evident: "are" and "is" should have been "were" and "was."[3]

Collins and Pinch acknowledge in the afterword to their second edition that "if the book is read, as it seems it can be, as an attempt to question the validity of relativity, then it gives a damaging and distorting account. That, however, is not what we intended" (1998a, 163). I am now persuaded that phrases like these were not designed to suggest to readers that the theory of relativity was insecurely based and regret having suggested that they were, because it distracted us from the real issues. Even then, however, I was not accusing Collins and Pinch of having, as they put it, "a bias against relativity" (1996, 11). Whether the theory pleases or displeases them is clearly of minor interest, but in fact I noted in Mermin 1996 that they described relativity as "beautiful, delightful, and astonishing" (Collins and Pinch 1993, 54). My complaint was not that they disapproved of the theory, but that their description of the accumulated evidence in its support during its first three decades was so incomplete as to suggest that its acceptance in 1933 was based on sparse and superficial evidence.

As their afterword now makes clear, Collins and Pinch did not intend to argue that relativity was accepted on flimsy grounds, but only that the folklore that has accumulated around those early decades gives a superficial account of that acceptance: "[W]hat we are trying to do is to show that most of what scientists have said to citizens about the founding of relativity is wrong" (1998a, 154). If "most" were changed to "much" I would go along with this conclusion. If the chapter in *The Golem* on relativity were clearly presented as nothing more than a debunking of two "crucial experiment" myths, I would have few problems with it.

But although Collins and Pinch (1997) have declared this to have been the chapter's primary purpose, it can easily be read as making stronger claims about the character of scientific knowledge. Because so little attention is given to any grounds other than these two experiments for taking special relativity seriously by the time of Dayton Miller's ill-fated 1933 claim to have found a small anisotropy in the velocity of light,[4] many readers will be given the impression that Miller's results

3. Harry, Trevor: That change didn't make it into the second edition. Perhaps the third?

4. Grounds for taking general relativity seriously at the time Collins and Pinch conclude their story were considerably weaker, and indeed, there were competing theories of gravitation.

were ignored only because in the quarter-century since Einstein's 1905 paper, physicists had simply acquired the habit of regarding the negative results of Michelson and Morley as definitive.

In a sense, of course, they had. But Collins and Pinch have little to say about what lay behind that habit. An enormous burden rests on one sentence: "Other tests of relativity . . . bolstered the idea that the theory of relativity was correct" (Collins and Pinch 1993, 42; Collins and Pinch 1998a, 42).[5] Unfortunately the sentence that follows this can be read to undermine the point: "The sheer momentum of the new way in which physics was done—the culture of life in the physics community—meant that Miller's experimental results were irrelevant." What this means depends, of course, on what "culture" is taken to include. If, as I now understand Collins and Pinch's intent, "the culture of life in the physics community" refers to the cumulative impact of all theoretical and experimental work bearing on relativity since 1905, then the sentence is correct. But their readers, who have been given hardly an inkling of the existence of these intricate and extensive features of the culture, are likely to conclude that the momentum was generated by nothing more than many years of growing more and more comfortable with the old negative Michelson-Morley result itself. The "culture of physics" would then appear to be little more than the prejudicial enshrinement of Michelson-Morley as an incontestable fact.

Collins and Pinch have responded to this criticism by emphasizing that much of the other evidence available at the time was also far from clear cut. But this does not render its existence irrelevant. Their omission of even a cursory survey of the abundant support for special relativity by 1933 is likely to lead readers to conclude wrongly that physicists were acting on quite thin evidence when they then rejected Dayton Miller's claims. "Without even a glimpse of the rest of the tapestry, I do not see how the lay reader can fail to conclude that relativity is fraudulent" (Mermin 1996b, 13).

Why did I make such a fuss about this? Collins and Pinch originally thought it was because I perceived them to have "defiled the holy of holies: relativity"[6] (Collins and Pinch 1996, 11) by calling attention to the very complex and ambiguous role the Michelson-Morley experiment played in its history. But it is precisely the discovery of such complexity and ambiguity, missing from the folklore, that makes the study of the

5. The second edition's deletion of the word "indirectly" is a step in the right direction.

6. As noted below, Barry Barnes (1998) had a similar response to my criticisms of his textbook.

history of physics so fascinating to its practitioners. Nothing is "defiled" by such tales. My criticism of how Collins and Pinch tell the tale is that while they now maintain that "[i]t is not our business to set out the grounds for the acceptance of relativity—a huge project for physicists and historians" (Collins and Pinch 1998a, 154), it ought to have been their business to make it clearer than they do that there is much more to the story than Michelson-Morley. From the case of relativity they draw the broad conclusion that the meaning of an experiment "depends on what people are ready to believe." True enough. But without an adequate sketch of what happened between 1905 and 1933 to affect what people were ready to believe, this can easily be read as a claim that purely subjective considerations play a dominant role.

Collins and Pinch ought to have sketched in the missing part of the story not out of reverence for a sacred text, but to avert a misreading of their chapter as a caricature of how scientific knowledge is constructed. My concern was not that they "were trying to give comfort to" a deviant subculture of "scientific outsiders" who to this day have not accepted relativity (1998a, 155). It was that they were likely to be misinforming the general public about its nature and, through the general conclusions they drew from the case of relativity, about the broader nature of scientific knowledge.

Theory versus Experiment

In their afterword Collins and Pinch call attention to the apparently paradoxical fact that "the comments of nearly everyone who has studied the period closely suggest that the decisive feature of special relativity that caused scientists to agree to agree in the way that they did about the experiments was the simplicity of structure that it brought to electromagnetic *theory*" (Collins and Pinch 1998, 167; emphasis in original). Indeed, in my column I noted that they left out of their story "the unity and coherence that relativity immediately brought to electrodynamics" (1996a, 11). My hunch (unbacked by historic research) is that this is what must have been the decisive point for those physicists, bold enough to grasp Einstein's extraordinary insight into the nature of time, who almost immediately embraced the special theory.

To be sure, one must add to this the clear failure of Michelson and Morley to find an unambiguously *positive* result (as opposed to the widely proclaimed, but highly problematic demonstration of an unambiguous *null* result). Had they found a convincing positive result, this would have made nonsense of Einstein's postulate that the principle of relativity must hold in electrodynamics as well as in mechanics. What was, I suspect, of crucial importance about Michelson-Morley was not its ability clearly to provide evidence in direct support of the theory, but its failure to provide clear-cut evidence that would have contradicted relativity from the outset. It is only in this negative sense and only after his remarks about electromagnetism that Einstein makes his passing allusion to such experiments as "mislungenen Versuche, eine Bewegung der Erde relativ zum 'Lichtmedium' zu konstatieren" [unsuccessful attempts to determine a motion of the Earth relative to the "light medium"] (1905, 891).

Collins and Pinch seem to be suggesting that because this early support for relativity came from the broad overall *theoretical* structure of electromagnetism, it is irrelevant to the question of the early *experimental* evidence in support of relativity. But if one pillar of the sociology of scientific knowledge (SSK) is that experiment is theory laden, surely another ought to be that theory is experiment laden. The fact that relativity brought coherence to electromagnetic theory makes the complex body of experimental lore encoded in that theory an important part of the experimental evidence supporting relativity. To bring a great simplification to electromagnetic theory is to impose a greater coherence on the body of experiments on which that theory rests.

Strands, Tapestries, and Ropes

In Mermin 1996a I said that in dismissing the additional unmentioned evidence as potentially problematic, Collins and Pinch overlooked the fact that "even though many clues in a complex network of evidence will always be far from definitive, the probability of a conclusion supported by a multitude of interlocking mutually reinforcing clues can still be close to certainty" (13). In Mermin 1996b I repeated this: "Hardly a hint is given that this essay follows a single tiny strand in an enormous tapestry of fact and analysis" (13). I stated it again in Mermin 1996c: "Michelson-Morley was only a small part of an intricate network of theory and experiment to which relativity brought clarity and coherence. That there may be a complex story, yet to be written, about any other purported 'crucial test' of relativity is beside the point; what the view of science in *The Golem* leaves out is that the existence of many different strands of evidence can transform a hypothesis into a fact, even in the absence of any single unimpeachable crucial experiment" (15).

Collins and Pinch picked up on this approvingly in Collins and Pinch 1997, remarking, however, that "strands of evidence can be woven in different ways. Thus, one still needs an explanation of why a group of scientists interprets a set of strands of evidence one way rather than another, and one needs to set this explanation in the context of an analysis that shows how different kinds of weaving could have been done" (92).

While the way in which the strands might be woven together is, of course, a complex and subtle matter, an adequate picture of the acceptance of relativity ought surely to mention their existence. My own feeling is that the notion that many different kinds of reweaving were readily available (the "Duhem-Quine thesis" of underdetermination) makes light of the daunting problem of how *plausibly* to weave together all the evidentiary strands in support of relavity by 1933 in any way other than the one we now accept. Duhem-Quine may be a neat logical point, but logic knows nothing of simplicity and gracefulness on the one hand and ungainly artificiality on the other.

In their afterword, Collins and Pinch reemphasize the probable weakness of the individual strands, anticipating that the many supporting threads that were not mentioned in *The Golem* will be found, after careful historical scrutiny, to have been significantly weaker than they are now retrospectively reconstructed to have been. Even granting this (and it does not strike me as unreasonable), there remains the centrally important fact of the strength of the rope, which is what my tapestry has turned into in their afterword. As they emphasize in their afterword, Collins and Pinch make much the same point in *The Golem:* "No test viewed on its own was decisive or clear cut, but taken together they acted as an overwhelming movement" (1993, 53). This would have greatly clarified the situation had they mentioned a representative sample of such tests. But they cite no tests whatever of special relativity other than Michelson-Morley itself, thereby giving their readers no adequate basis for judging how strong the full collection of strands might have been [16].

We nevertheless now appear to be in considerable agreement on this point. I suspect the major remaining bone of contention is not the weakness of many of the threads nor the strength of the tapestry-rope, but the uniqueness of the weaving, which, Duhem-Quine to the contrary notwithstanding, becomes, in my experience, more and more tightly constrained, as more and more strands need to be added.

Repeating What One Has Been Told

In Mermin 1996b I remarked that even though Collins and Pinch "state at the end that 'we have no reason to think that relativity is anything but the truth'" on the basis of the history they present in *The Golem,* a reader "can only wonder why they might think so" (13). In their afterword Collins and Pinch say that never having done any experiments bearing on the validity of relativity themselves, "when our critics berate us for not endorsing the truth of relativity sufficiently vigorously, or for seeming to cast doubt on the truth of relativity by our historical treatment, they are really demanding that we endorse what we have been told" (1998, 173).

But I was not criticizing a failure of Collins and Pinch to affirm with appropriate vigor their confidence in a body of knowledge that they had no direct competence to assess. My criticism was of their inadequate acknowledgment of the existence and relevance of such a body of knowledge. No endorsement by Collins and Pinch of the validity of that knowledge would have been required or, as they correctly note, sensible.

Behind this miscommunication lies an important distinction between collective and individual knowledge that ought always to be made explicit.[7] When most people speak of "scientific knowledge," they mean collective knowledge, extending far beyond the direct experience of any single person. I have never been to China. I can nevertheless assure people that such a place exists, because the conspiracy necessary to fool me into believing it, were it false, is too implausible to contemplate. I could cite evidence from atlases, encyclopedias, and traveling friends to support this point. For me this evidence is entirely social, but it is evidence, nonetheless. It is no different with relativity. Even though Collins and Pinch have done no experiment themselves—nor have I—there is much they could legitimately have said to give a more accurate picture of the basis for our collective belief that it is a valid body of knowledge.

Stepping into the Shoes of Others

Can present knowledge be relevant to a proper understanding of how knowledge was constructed in the past? In their afterword Collins and Pinch state that their approach to the history of science is to get "into the shoes of the scientists, sharing only the knowledge that the scientists could have had at that time." To do otherwise, they maintain, would give "an impression of simplicity and success which misleads when we are faced with the scientific and technological dilemmas that are found

7. Harry Collins and I slowly realized in the course of several e-mail exchanges that by "our knowledge" I always had in mind collective knowledge, but he was always thinking of the knowledge of any given individual. Realizing this eliminated much unnecessary misunderstanding.

in contemporary life." To make use of present knowledge would be "damaging to those who need to understand, not the contents of science, but the way scientific facts are established" (1998a, 166–67).

The sociologist or historian clearly must distinguish what was known at the time under study from what is currently known. It is all too easy for late-twentieth-century knowledge to induce anachronistic misinterpretations of how knowledge developed in the early twentieth century. All the historic folklore physicists tell each other suffers from such distortion. But preventing such contamination by prohibiting any use of or reference to contemporary knowledge has its own risks. By imposing a cut-off date on a study and playing the game that they must be time travelers back to the period with total amnesia about all contemporary knowledge, sociologists can protect themselves from giving trivial or circular answers to their questions, at the risk of depriving themselves of important clues about what kinds of questions it might be interesting to ask.

To understand why special relativity was so much a part of the culture of physics by 1933 that reported observations of an ether drift could simply be shrugged off, it would have been helpful to have had in mind present reasons for believing the theory, to help guide an inquiry into whether plausible versions of this more extensive evidence were available as early as 1933. Today, for example, relativity is widely viewed as having provided the final finishing step in the electrodynamic synthesis: the unification of electricity and magnetism cries out for a subtle revision of the concepts of space and time in order to achieve its most coherent expression. But this was also clear to Einstein from the start, and it surely had much to do with the early acceptance of the theory. A properly informed historian or sociologist ought to be on the alert for manifestations of this *Weltanschauung* in the early days, and not leave it out of "what you ought to know" about the acceptance of relativity.

Today relativity has achieved spectacular confirmation in its application at the atomic and subatomic scale, where it is crucial for a proper quantitative understanding of the dynamical behavior of electrons, protons, neutrons, and the more esoteric kinds of elementary particles. While the situation in particle physics was far from clear cut in 1933, quantum mechanics being less than a decade old, there were already at that time strong signs that relativity was an essential tool in accounting for the interaction between matter and light at the atomic scale. While it would indeed be absurd to use the triumphs of relativistic quantum electrodynamics in the 1950s to explain the indifference to Miller's results in the 1930s, there is nothing circular in using the knowledge of that triumph as a clue that the acceptance of relativity in 1933 might well have hinged, in part, on its uses in the atomic and subatomic physics of that period.

Schoolchildren and Savants

People have criticized the parable at the end of *The Golem,* where the emergence of a consensus among laboratory scientists is compared with the emergence of a consensus in a classroom of schoolchildren. Agreement is imposed by the teacher who assures the kids, who have each measured the boiling point of water and obtained readings all over the map, that they have just demonstrated that water boils at exactly 100 degrees Celsius (Collins and Pinch 1993, 150–51). In their afterword to the second edition, Collins and Pinch take these critics to be upset by the disrespect shown eminent scientists by the comparison to sloppy schoolchildren (1998a, 176–77).

But the problem with the parable is not that it fails to accord great scientists due reverence. The problem is that it suggests that the construction of scientific knowledge is nothing but the imposition of a preconceived answer by a figure of authority. Were I Collins and Pinch, I would have addressed this criticism by emphasizing that what was relevant in their parable was not the authoritarian way in which order was imposed on the muddled set of results, but the existence of the muddle itself, at an earlier stage of the process. They were offering a critique of the kind of science education that focuses exclusively on the consensus and has little if anything to say about the confusion from which the consensus emerges.

But parables can take on a life of their own. The picture of how the consensus is negotiated—by one-sided dicta from the teacher and in the absence of any attempts to repeat the measurements with greater care to discover possible sources of error—might strike Collins and Pinch as an incidental part of the story. But for most scientists it leaves out the heart of the process. The impression the parable leaves, of a superficial rush to a preconceived conclusion, is enforced by Collins and Pinch characterizing the classroom negotiation as illustrative of "the tricks of professional frontier science" (1993, 151). To be sure, their claim is only that their parable illustrates the scientific process better than the myth that laboratories are engines for the mechanical extraction of unambiguous truth. But whether or not their caricature is preferable to the caricature behind the "neat and tidy scientific myth," it remains a caricature. Changing "scientific myth" to "methodological myth" (as they have done in their second edition [1998a, 149]) does not get around its shortcomings.

I agree with Collins and Pinch that the myth of neat and tidy science ought to be challenged. But to be effective, a challenge should not set up a countermyth that is just as wide of the mark. Somewhere between the extremes of a well-oiled machine systematically generating universal truths and a chaotic classroom where anything goes and the answers are decreed by the authorities, there is a more nuanced view of the scientific process that acknowledges both the mess and the intellectually coherent process by which it is eventually cleaned up.

A SUBSEQUENT EXCHANGE WITH BARRY BARNES AND DAVID BLOOR

An exchange I had with two of the authors of *Scientific Knowledge* (Barnes, Bloor, and Henry 1996) resonates with my conversations with Collins and Pinch. I thought I had learned from my earlier experience how to avoid elementary misunderstanding arising out of differences in style or rhetoric, and indeed, things could have been much worse. Nevertheless, many of the kinds of talking past each other that characterized my earliest interactions with Collins and Pinch were strikingly repeated.

Priests and Astrologers

Collins and Pinch originally found in my two columns "the whiff of the offended priest": "[O]ur having defiled the holy of holies: relativity . . . has touched a raw nerve" (1996, 11). In a similar vein, Barry Barnes began his reply to my criticism of *Scientific Knowledge* with the observation that "people who have an intense respect for something . . . often feel that to speak about it threatens to demean and destroy it—that speech of this kind is a pollution. . . . Increasingly it is felt necessary to reassure everyone . . . just how extraordinarily reliable science and scientific knowledge are. . . . [I]f you value something you had best jump up and down and drool about how wonderful it is" (Barnes 1998, 636–37).

This outburst seems to have been provoked by my insistence that there were more instructive examples than astrology to choose as a textbook example of an area defined to be outside the boundaries of science that could conceivably shift back in. Astrology struck me as a bad example because a reconstruction of scientific knowledge to accommodate the knowledge claims of astrology would have to be massive and far-reaching on an unprecedented scale. Probably, though, I was gratuitously provocative when I remarked that "undermining science" did not strike me as an unreasonable way to view the selection of that particular example for use in an introductory text on the sociology of scientific knowledge (1998a, 636).[8]

To my disappointment, Barnes read my suggestion that astrology was a bad example as an expression of "doubt as to whether the authors of *Scientific Knowledge* might perhaps be into astrology." With heavy irony he reassured me that he has "no time for horoscopes," although, he added, "the belief state of an obscure lump of molecules wandering around a remote corner of England is of no importance whatsoever epistemologically speaking" (1998, 637). Compare this with Collins and Pinch's "Let us now state that we have no bias against relativity" (1996, 11), followed by the remark that it really doesn't matter whether they—Harry Collins and Trevor Pinch—"believe in relativity."

In both cases the response was not to what I was saying, but to why I might have been saying it. Both times I was taken to be responding to a perceived violation of something I held sacred, and my actual criticism—that they were paying insufficient attention to the broad coherence of an extensive body of knowledge—was read as a charge that they personally were biased (against relativity, for astrology). In both cases the imagined charge was both denied and dismissed as irrelevant, while the substance of my criticism was not addressed.

Subsequently Collins and Pinch and I learned how to talk to each other. Our discussion turned to matters of substance in our talk about weaving, whether into tapestries or ropes. Bloor's response also opened the way to further discussion. He noted that by dismissing as casually as I did purported evidence in support of astrology I was "taking for granted the acceptability of current forms of understanding" (1998, 626), which was just the point they were trying to make with their astrological example. I find "taking for granted" an odd way to characterize an unwillingness to invest a major effort in the investigation of grossly improbable claims, but in contrast to Barnes's response, Bloor's invites further discussion.

The moral for a better level of dialogue seems clear. Each must attend more carefully to what the other is actually saying, resisting the impulse to infer and respond only to a supposed motivation for saying it, or the manner in which it seems to be being said. I succumbed to this tempta-

8. I return below to why I made the remark.

tion in my columns on *The Golem* when I suggested that Collins and Pinch were trying to induce skepticism in their readers by various rhetorical tricks. I thereby emitted a distracting priestly whiff. I got into similar trouble with Barnes and Bloor by scattering through my critique responses to a remark in their introduction that some scientists "have assumed that since we neither praise nor defend science our objective must be to subvert it" (1996, viii). This reading of their scientist critics struck me as quite wrong, so whenever I came upon passages that could reasonably be misconstrued as hostile to science or scientists, I called attention to them. I was not trying to make a case that the authors were, in fact, hostile, but I did want to give them a sense of why some readers might read them not as damning with faint praise, but simply as damning. But all I conveyed to Barnes was that I was "obsessed with the search for non-existent 'subversion'" (Barnes 1998, 639). Fair enough. I should have taken greater pains to explain what I was up to.

Unlicensed Scientists?

The other striking similarity in the responses both of Collins and Pinch and of Barnes and Bloor was that they all read me as charging sociologists with trying to do physics themselves, and doing it badly—practicing science without a license. Collins and Pinch disavowed the notion that they, a pair of sociologists—not physicists—could have been moved to mount a challenge to physics. "Sociologists are not physicists, and it is [not] their business to offer opinions about the findings of physics" (1997, 92). In a very similar vein, Bloor talked about my "failure to see that the sociologist of physics isn't doing physics and the sociologist of biology isn't doing biology. . . . [T]here is [in Mermin] a suggestion that a stance within the sociology of physics or biology will be a stance within physics or biology itself, and hence must be monitored to see if it is good physics or biology" (1998, 625).

As an example of this Bloor said that I took "what we have to say about the concept of carbon to be a challenge to the current physics of carbon" (1998, 625). Barnes and Bloor use the case of carbon to illustrate their claim that every act of classification is defeasible and revisable—that no act of classification is ever indefeasibly correct. One of my main criticisms is that they base this very general conclusion on a theory of classification that deals only with continuously variable categories, never mentioning any of the very important classifications based on discrete differences (like those between squares and triangles). In the case of carbon, they cite the problematic character of a category that embraces both diamonds and soot, without saying a word about the underlying atomic structure—the basis for a discrete classification scheme—that accounts for diamond and soot being viewed as different forms of a single entity, carbon.

An adequate defense of this theory of the malleability of classifications requires an analysis of cases where the classification rests on differences that are discrete, and not continuously variable. I was not criticizing Barnes and Bloor for *challenging* the grounds for regarding diamonds and soot as both consisting of carbon. I was criticizing them for *failing to mention* those grounds, which are not accommodated by their general theory of classification. In much the same way, I criticized Collins and Pinch not for challenging special relativity, but for failing to mention any evidence in support of it by 1933 other than Michelson-Morley.

But I put my point in a way that Barnes and Bloor could misread as a claim that they were promulgating new physics of dubious merit. I said that a "successful challenge to the idea that there is something unmistakably 'carbon' of which many diverse materials are simply different geometric arrangements, would require one to restructure a vast body of knowledge from chemistry, solid-state physics, biology, nuclear physics, and even astrophysics" (1998a, 614). Barnes and Bloor took this to refer to a (nonexistent) challenge by them. What I meant, though, was a hypothetical challenge by physicists attempting to take advantage of the alleged flexibility of classifications to reconstruct the category of carbon with diamond on one side of the boundary and soot on the other. This was part of my broader suggestion that SSK exaggerates the interpretive flexibility available to scientists trying to construct a coherent and internally consistent body of knowledge.

I had expected Barnes and Bloor to reply with an argument that the categories I regarded as discrete were actually based on subtle kinds of continuous variability that I had overlooked—that the discreteness was, in some interesting way, illusory. I think this would have been a hard case to make, but I anticipated that they would make a serious attempt at it. Without such an argument, it seems to me their theory of classification is irrelevant to much of science as it is actually practiced.

Encouragingly, David Bloor did not misunderstand me to anything like the extent that Barry Barnes did. While Bloor did misconstrue my remarks about the inaccuracy or incompleteness of their descriptions of scientific practice as claims that they were advocating new scientific theories of their own, he viewed this as conceptual confusion on my part. He did not see me as Barnes did—as wanting only to "protect the external image of his subject" or to "protect [the general reader] *with* oversimplifications from truths that might otherwise do her harm," or as taking the position "that what he, a scientist, is able to approve of cannot equally be approved of by others" (1998, 639).

Perhaps Barnes simply read me less carefully than Bloor did. A friend remarked that Barnes's reply to my article was really quite eloquent and compelling as a reply to many criticisms of SSK by scientists—but not to the criticism he was supposed to be addressing. But I suspect the difference between Barnes's response and Bloor's has more to do with the fact that thanks to Harry Collins, David Bloor and I had spent three days together the summer before our exchange. At the end of that long weekend it was impossible for me to believe that his primary aim was to undermine the public image of science and impossible (I hope) for him to believe that mine was to exalt it.

A FEW SIMPLE CONCLUSIONS

These histories offer some elementary lessons in how scientists and sociologists should converse, not in their roles as anthropologists and native informants, but as academic colleagues, reflecting on the nature of their two disciplines.

Rule 1: Focus on the substance of what is being said and not on alleged motives for saying it.

Some scientists are indeed devoted to worshipfully polishing the public image of science; some sociologists do indeed enjoy puncturing such celebratory balloons. But one must beware of the powerful impulse to infer and become distracted by such behavior when none is intended. Even when glorifying or debunking seems indisputably behind something you disagree with, it is not a valid refutation to point this out and stop, as if the mere existence of a desire to elicit reverence or scorn were enough to invalidate the argument being put forth for that apparent purpose.

Rule 2: Do not expect people from remote disciplines to speak clearly in or understand the nuances of your own disciplinary language.

It took me and Harry Collins far too long to realize that for him "knowledge" without qualification meant the personal knowledge of a single individual, while for me it meant the collective knowledge of a large group. Until this discovery each of us found much of what the other had to say about the relation of knowledge to "what I have been told" quite preposterous, and our responses to each other only heightened the sense of absurdity. "Negotiations" is another tricky term. For sociologists it is a morally neutral characterization of the process by which different people come to a mutually acceptable understanding. But for most scientists it has overtones of duplicity and personal self-interest and suggests cynically splitting the difference in a disagreement rather than searching for a deeper understanding.

Rule 3: Do not assume that it is as easy as it may appear for you to penetrate the disciplinary language of others.

This is rule 2 again, viewed from the other direction. For while it may be obvious to you that others are missing your point, it requires a much greater expenditure of creative imagination to contemplate the possibility that you might be missing theirs. In particular, it is necessary to seek for interpretations of what they are saying that are less absurd than the one that first crosses your mind. Collins and Pinch, and Barnes and Bloor, should have thought again before concluding that I was doing anything as silly as reading them to be putting forth unorthodox physics of their own devising. I should have asked myself whether Collins and Pinch could really have been engaged in a quixotic effort to undermine public confidence in relativity. You cannot argue successfully with people without first persuading them that you have more or less understood what they are trying to say.

In short, assume that people in another discipline really mean what they say, take the trouble to ascertain just what that might be, and, when you are confident of this, take care to formulate your criticism, if you still feel criticism is called for, in terms you can be confident will be understood. These rules for fruitful conversation are so obvious that I feel foolish setting them down. But their neglect, on both sides, is responsible for much of the heat of the science wars.

Chapter 8

HOW TO BE ANTISCIENTIFIC

Steven Shapin

I am not a commissioned officer in the so-called science wars. If anything, I am something between a common soldier and an interested witness to the current hostilities. I was trained in genetics, but for many years I have been a historian and sociologist of science, writing mostly about the development of science in the seventeenth century.[1] I have suffered some minor shrapnel wounds from wildly aimed shells, but, in the main, the Defenders of Science have had bigger game to stalk and have left me to get on with my work and to reflect from a somewhat disengaged perspective on what is going on.

The immediate occasion for the science wars seems to be a series of claims *about* science made by some sociologists, cultural historians, and fuzzy-minded philosophers. (In my ordinary academic work, distinctions between these categories—and subdivisions within them—count as crucial, but in this piece for general readers I mainly lump them together.) As a matter of convenience, I refer to propositions *about* science as "metascience," and, because it is very important to be clear about what is at issue, I list here just a few of the more contentious and provocative metascientific claims:

1. There is no such thing as the Scientific Method.
2. Modern science lives only in the day and for the day; it resembles

A French translation of a slightly different version of this essay has already appeared as "Etre ou ne pas être antiscientifique," *La recherche* 319 (April 1999), 72–79, and an excerpt has been published in German as "Von der Schwierigkeit, ein Wissenschaftsgegner zu sein," *Frankfurter Rundschau,* 27 October 1998, sec. Humanwissenschaften, 9.

1. Some of my work in this area includes Shapin and Schaffer 1985 and Shapin 1994, 1995a, 1995b, 1999, and forthcoming.

much more a stock-market speculation than a search for the truth about nature.

3. New knowledge is not science until it is made social.

4. An independent reality in the ordinary physical sense can neither be ascribed to the phenomena nor to the agencies of observation.

5. The conceptual basis of physics is a free invention of the human mind.

6. Scientists do not find order in nature, they put it there.

7. Science does not deserve the reputation it has so widely gained . . . of being wholly objective.

8. The picture of the scientist as a man with an open mind, someone who weighs the evidence for and against, is a lot of baloney.

9. Modern physics is based on some intrinsic acts of faith.

10. The scientific community is tolerant of unsubstantiated just-so stories.

11. At any historical moment, what pass as acceptable scientific explanations have both social determinants and social functions.

For many readers, even listing such statements is unnecessary: they will already be thoroughly familiar with sentiments like these associated with the writings of sociologists of science and academic fellow-travelers, as they will be equally familiar with the outraged reactions to them expressed by a number of natural scientists, convinced that such claims are motivated mainly or solely by hostility to science, or that they proceed from ignorance of science, or both. Science and rationality are said to be besieged by barbarians at the gate, and, unless such assertions are exposed for the rubbish they are, the institution of science, and its justified standing in modern culture, will be at risk. It is therefore incumbent on leading scientists themselves to speak out, to say what the real nature of science is, and to take a stand against the ignorance and the malevolence expressed in these claims.[2]

Nevertheless, I have to tell you—in the spirit of our troubled culture—that you have just become a victim of yet another hoax. None of these claims about the nature of science that I have just quoted, or minimally paraphrased, does in fact come from a sociologist, or a cultural studies academic, or a feminist or Marxist theoretician. Each is taken from the metascientific pronouncements of distinguished twentieth-century scientists, some Nobel Prize winners. (See the end of

2. Some well-known recent tracts by scientists expressing such sentiments are Wolpert 1992, Gross and Levitt 1994, Gross, Levitt, and Lewis 1996, Sokal and Bricmont 1998, and Weinberg 1995 and 1998.

this chapter for a list of the sources.) Their authors include immunologist Peter Medawar, biochemists Erwin Chargaff and Gunther Stent, entomologist E. O. Wilson, mathematician turned scientific administrator Warren Weaver, physicists Niels Bohr, Brian Petley, and Albert Einstein, and evolutionary geneticist Richard C. Lewontin. This is not a mere a party trick—a device to turn the tables or to play intellectual Ping-Pong—though it would seem so if I left it at that. The point I want to make here is substantial, interesting, and potentially constructive: practically all of the claims about the nature of science that have occasioned such violent reaction on the part of some recent Defenders of Science have been intermittently but repeatedly expressed by scientists themselves: by many scientists of many disciplines, over many years, and in many contexts [14].[3]

Accordingly, we can be clear about one thing: it cannot be the claims themselves that are at issue, or the claims themselves that must proceed from ignorance or hostility. Rather, it is *who has made* such claims, and what motives can be attributed—plausibly, if often inaccurately and unfairly—to the *kinds of people* making the claims. So one of the very few, and very minor, modifications I have made in several of the quotations above is the substitution of the third-person "they" or "scientists" or "physicists" for the original "we." We are now, it seems, on the familiar terrain of everyday life: members of a family are permitted to say things about family affairs that outsiders are not allowed to say. It is not just a matter of truth or accuracy; it is a matter of decorum. Certain kinds of description will be heard as unwarranted criticism if they come from those thought to lack the moral or intellectual rights to make them.

Since what scientific family members often do when they make metascientific statements is to *prescribe* how members *ought* to behave—criticizing or praising—there is a tendency to assume that outsiders must be about the same business, though without equivalent entitlements. It is sometimes hard for scientists to understand how the description and interpretation of science could be anything other than coded prescription or evaluation: telling scientists what to do, or sorting out good from bad science, or saying that science as a whole is good or bad. It is hard to recognize, that is, what a naturalistic intention would be like in talking about science, since this is not a luxury readily available to members of the scientific family. Scientists have naturalistic inten-

3. After I had written this essay, I came across a broadly similar observation powerfully made by Israeli historian of physics Mara Beller (1998; see also Beller 1997), though she focuses attention exclusively on the views of twentieth-century quantum physicists.

tions with respect to their objects of study but rarely with respect to the practices for studying those objects. So, for example, some sociologists do indeed insist that scientific representations are "social constructions." And when some scientists read this they assume—wrongly, in most cases and in my view—that these sociologists have tacitly prefaced the phrase with the evaluative word "only," or "merely," or "just": science is *only* a social construction. To say that science is socially constructed is then taken as a way of detracting from the value of scientific propositions, denying that they are reliably about the natural world.[4] Scientists do that all the time: that is, they "deconstruct" particular scientific claims in their fields by identifying them as *mere* wish-fulfillment, *mere* fashion, *mere* social construction. But they do so to *do* science, to sort out truth from falsity about the bits of the natural world with which they are concerned. They rarely do so with what might be called a disciplinary intention of just describing and interpreting the nature of science. That is one major reason why we seem to be misunderstanding each other so badly. There are important differences in recognized disciplinary intentions, in seeing their different possibilities and purposes and values. We do not always adequately recognize these differences, and we ought to.

That is one lesson to take away from this little hoax. But it is neither the most interesting nor the most fundamental. The more fundamental observation is just that metascientific statements by scientists vary enormously. I have picked out some that resonate with descriptions offered by sociologists, but, of course, there are many that do not. When scientists say metascientific things, they commonly conflict with each other as well as conflicting, occasionally, with what sociologists say.

Indeed, some scientists' pronouncements on the nature of science insist that science is a realist enterprise; others stipulate that it is not. Science, these others say, is a phenomenological, instrumental, pragmatic, or conventional practice. Max Planck, for example, identified the endemic tendency "to postulate the existence of a *real world,*" in the metaphysical sense, as "constitut[ing] the irrational element which exact

4. Sociologists of science, notably those Edinburgh school writers criticized by Steven Weinberg and others, have repeatedly stressed that the social component of scientific knowledge is *not* to be set against the causal role of unverbalized natural reality: the social component is seen as a condition for having experience of a recognized kind and for representing that experience in linguistic form. See, for example, Bloor 1991: "No consistent sociology could ever present knowledge as a fantasy unconnected with our experience of the material world around us" (33) and Barnes 1977: "[T]here is indeed one world, one reality, 'out there,' the source of all our perceptions" (25–26); see also Barnes 1992. I have no very satisfactory ideas why the Defenders of Science should miss the facts right in front of their eyes.

science can never shake off, and the proud name, 'Exact Science,' must not be permitted to cause anybody to under-estimate the significance of this element of irrationality" (1949, 106). J. Robert Oppenheimer supposed that laypeople were irritated by scientists' *un*willingness to use words like "real" or "ultimate": the use of such notions would be a form of metaphysics, and science, Oppenheimer insisted, was a "non-metaphysical activity" (1954, 4). These positions are hard to square with such nervously defiant declarations as Steven Weinberg's: "[F]or me as a physicist the laws of nature are real in the same sense (whatever that is) as the rocks on the ground" (1998, 52).[5] As it happens, physicists disagree on such things.

Moreover, some scientists—when they say that science is a realist enterprise—mean to pick out a special philosophical position by which theoretical entities are understood to refer to real existents in the world; others seem to be alluding to the sort of robust everyday realism that unites a range of sciences with the practices of everyday life, as when I might say in ordinary conversation, "Look at the cat sitting on the mat," directing someone's attention *over there* and not toward my speech organs or my brain. The realism advocated (or rejected) in scientists' metascientific pronouncements is only very occasionally specified in such ways. Some scientists say that science aims at, or arrives at, one universal Truth, others say that the truths of sciences are plural, or that science is just "what works" and that Truth, or even correspondence with the world, is none of their concern—just "what is the case" or "what seems to be the case to the best of our current efforts and beliefs." Some say that science is Coming to an End—about to be completed—but we should understand that this imminent completion has been promised practically as long as there has been science. Other scientists pour scorn on any such idea: science, they say, is an open-ended problem-solving enterprise, where the problems are generated by our own current solutions and will continue to be, time without end.[6]

Some scientists' metascientific pronouncements say that there is no such thing as a special, formalized, and universally applicable Scientific Method; others insist with equal vigor that there is. The latter, how-

5. Only after this piece was drafted did I become aware of Richard Rorty's similar, but more vigorously expressed, puzzlement about Weinberg's claim (Rorty 1997).

6. The controversy among scientists about whether or not science is about to be completed now even claims space in the *New York Times:* see the debate between John Horgan and John Maddox (1998). For pertinent claims, see Weinberg 1992; Horgan 1996; Horgan and Maddox 1998; and Stent 1969. For historical commentary on recurrent announcements of the End of Science, see Schaffer 1991. For my own engagements with what scientists have meant by truth, see Shapin 1999 and forthcoming.

ever, vary greatly when it comes to saying what that method is. Some scientists like Bacon, some like Descartes; some go for inductivism, some go for deductivism; some for hypothetico-deductivism, some for hypothetico-inductivism. Some say—with T. H. Huxley, Max Planck, Albert Einstein, and many others—that scientific thinking is a form of common sense and ordinary inference. "The whole of science," according to Einstein, "is nothing more than a refinement of everyday thinking" (1954, 319).[7] Others, like the biologist Lewis Wolpert (1992), vehemently repudiate the commonsense nature of science and suggest that any such idea stems from ignorance or hostility. Few—either for or against the commonsense nature of science—display much curiosity about what common sense is or entertain the possibility that it too might be heterogeneous and protean.

You name it, it's been identified as the Scientific Method, or at least as the method of some practice anointed as the Queen of the Sciences, the most authentically scientific of sciences—usually, but not invariably, some particular version of modern physics. Collect textbook statements about the Scientific Method and see for yourself. Or ask your scientist-friends, one by one, to write down on a piece of paper (no collaborating! no peeking at a philosophy of science textbook!) what they take to be either the Scientific Method or even the formal method thought to be at work in their own practices or discipline. Some of your friends will have heard of Karl Popper, or of Thomas Kuhn, or of Paul Feyerabend and will have their preferences among these—though probably not many of them. (Why should they?) In which case, ask them to write down on another piece of paper what they take to be the position about Scientific Method recommended by their favorite philosopher. (You may find little correspondence with sociologists' or philosophers' professional sense of what Popperianism or Kuhnianism is, and, in any case, sociologists and philosophers also vary in their estimation of what Popper and Kuhn were really saying.)[8]

You might also consider the cultural sources of our current repertoires for talking about Scientific Method. Few chemists, biologists, or physicists will have taken courses on Scientific Method (at least in Anglophone settings), but many psychologists or sociologists will have experienced almost total immersion in such material—ironically taken

7. For Huxley, see Huxley 1900: "Science is, I believe, nothing but *trained and organised common sense*" (45); for Planck, see Planck 1949, 88.

8. For an interesting exploration of what scientists' professions of Popperianism might mean, see Mulkay and Gilbert 1981; for psychological assessments of scientists' grasp of formal logic, see Mahoney 1979 and Mahoney and DeMonbreun 1977.

to be modeled on formal natural scientific method. Perhaps no small part of the enormous success of the natural sciences might be ascribed to the relative *weakness* of formal methodological discipline [18]. It is at least a thought worth thinking. This was, for instance, the opinion of the physicist Percy Bridgman: "It seems to me that there is a great deal of ballyhoo about scientific method. I venture to think that the people who talk most about it are the people who do least about it. Scientific method is what working scientists do, not what other people or even they themselves may say about it. No working scientist, when he plans an experiment in the laboratory, asks himself whether he is being properly scientific, nor is he interested in whatever method he may be using *as method*. . . . The working scientist is always too much concerned with getting down to brass tacks to be willing to spend his time on generalities. . . . Scientific method is something talked about by people standing on the outside and wondering how the scientist manages to do it" (1955, 81).

When we consider the *conceptual* identity of science, the situation is much the same. Is science conceptually unified? To those scientists who consider that it is, a preferred idiom is a unifying materialist reductionism, though scientists of a mathematical or structural turn of mind reject both materialism and reductionism, while biologists continue intermittently to ponder whether there is a not a unique biological mode of thinking and unique biological levels of analysis. Just as E. O. Wilson is announcing a new—or rather a revived—plan for the reductionist unification of the sciences, natural and human, other scientists rebel against reductionism, against the claim that "the whole is the sum of its parts," or against its local manifestations in molecular biology, or they say that what had once been a search for understanding has now turned into a reductionist and shallow quest for explanations. Materialistic reductionism is just a sign that a Scientific Age of Iron has followed an intellectual Golden Age.[9]

The conceptual unification of all the sciences on a hard and rigorous base of materialist reductionism is an old aspiration, but it has never

9. For the most aggressive recent affirmation of reductionist unity, see Wilson 1998b, though Wilson now seems to have forgotten the complaints against rampant molecular reductionism he so eloquently expressed in his autobiography *Naturalist* (1995, chap. 12). Violently antireductionist statements by biologists are not, of course, hard to find: see, among many examples, Shulman 1998, Mayr 1997, Chargaff 1963 and 1978, and Lewontin 1993. For what it is worth, Wilson's vision of reductionist unity is devastatingly taken apart by the philosopher Jerry Fodor: "[Wilson] suspects that if we resist consilience, that's because we're suffering from pluralism, nihilism, solipsism, relativism, idealism, deconstructionism and other symptoms of the French disease" (1998, 3, 6).

commanded (and does not now command) the assent of all scientists. In a whole range of natural sciences—though biology is probably the most pertinent case—reductionist unification is rejected, sometimes very violently, and in other parts of science reductionist unification just doesn't figure. It may be somebody's dream, but it's hardly anybody's work.

Recall that I started by picking out claims about the nature of science that I invited you to associate with ignorant or hostile nonscientists. Then I told you that these statements were in fact made by scientists. Taking the argument a step further, I then acknowledged that metascientific statements by scientists were very various—on all subjects, and on all levels—and that many of these conflicted with sentiments in the quoted set, and with each other.

From this circumstance one could draw a number of conclusions. The first would be that a certain set of these statements—say the first set—is hopelessly in error and that their opposites are correct. I don't want to say that. If I did, it would be as much as saying that Medawar, Planck, and Einstein didn't know what they were talking about, nor do the sociologists whose claims resemble theirs so closely. In all honesty, however, I have to admit that when I plow through the range of individual scientists' metascientific statements I often find more internal variability than makes me professionally comfortable. I might even be accused of the sin of quoting isolated remarks out of context, and maybe I have. No one should tendentiously quote out of context, though perhaps quoting Peter Medawar out of context on the Scientific Method is a less serious offense than (I take a randomly chosen example) quoting Steven Shapin out of context on the role of trust in seventeenth-century English science: Medawar's proper business is less damaged by such misleadingly selective quotation than is mine. It is bad to quote out of context, or to quote misleadingly. It is bad for sociologists to do when writing about science or metascience, and it is bad for scientists to do when writing about the sociology of science. No, I want to say that the quoted set contains quite a lot of truth—with some qualifications that I am shortly going to make.

The second conclusion would be that all metascientific statements by practicing scientists are best ignored. For this view—at the risk of introducing a Cretan paradox—I can cite prominent scientists' pronouncements too. It was, after all, Einstein who famously said that we should take little heed of scientists' formal reflections on what they do; we should instead "fix [our] attention on their deeds": "It has often been said, and certainly not without justification, that the man of science is a poor philosopher" (1954, 296, 318).[10] So, if we follow Einstein and charitably allow the self-contradiction to pass, what one would be tempted to say is something like this: "Plants photosynthesize; plant biochemists are experts in knowing how plants photosynthesize; reflective and informed students of science are experts in knowing how plant biochemists know how plants photosynthesize."[11] As Aesop put it, the centipede does marvelously well in coordinating the movements of its hundred legs, less well in giving an account of how it does so. No skin off the centipede's back, and no skin off the scientist's back if it happens that she's not very good at the systematic reflective understanding of her work. That's not her job. And the point, of course, of Aesop's fable is that the centipede pushed to reflective understanding winds up in an uncoordinated heap. Kuhn just follows Aesop in this regard.

That's not really the conclusion I want to press either, though it does have something to recommend it. I see no necessary reason why certain scientists—perhaps not very many, given the pressures on their time and their other interests—shouldn't be just as good at metascience as professional metascientists, nor any necessary reason why professionals in metascience should ignore the pronouncements of amateurs. Nor do professional metascientists—sociologists, historians, and philosophers—globally *have to* concede that practicing scientists "know the science better or best" or "know more science" than they themselves do, though it is very prudent to respect scientists' particular expertise and to make sure, when one is writing about the object of that expertise, to "get it right." They should take great care not to say something about photosynthesis or about the techniques for knowing about photosynthesis that is demonstrably wrong, as judged by the consensus of expert practitioners in that area.

The reason that sociologists, historians, and philosophers do not globally have to concede that "scientists know better about science" is that knowledge about contemporary plant biochemistry, for instance, is not the same thing as "knowledge about science." There are many sciences at time present, and there have been many more sciences, and many versions of plant science, in past times, and who is to say that the historian or sociologist who knows something substantial about these many sciences knows "less science" than the contemporary plant bio-

10. Quoted more fully, Einstein said: "If you want to find out anything from the theoretical physicists about the methods they use, I advise you to stick closely to one principle: don't listen to their words, fix your attention on their deeds" (1954, 296).

11. I believe I owe this formulation to a conversation with Harry Collins many years ago.

chemist who, pronouncing on the nature of science, knows less or even nothing at all?

I see no reason to turn the tables and celebrate as a fact that I know "more science" than my friend who is a plant biochemist. As it happens, I know almost nothing about photosynthesis beyond what I was taught in college courses in plant physiology and cell biology, and I would be morally wrong and intellectually careless if I pronounced on how matters stand in that part of present-day science. On the other hand, I have the right to feel slightly miffed if I am lectured about how matters stood in seventeenth-century pneumatic chemistry by practicing scientists who are even more incompetent in that part of science than I am in contemporary plant biochemistry.

Almost needless to say, it's vital that you get your facts right in the subject you're writing about. That obligation is absolute and it's general: it applies to sociologists and historians writing about the aspects of science in which they are interested, and it applies to scientists writing about the sociology and history of science. At the same time, one would hope that normal human and professional frailties would be recognized and that we would pause a nanosecond before ascribing to each other the basest possible motives and the most egregious degrees of incompetence. There is indeed some shoddy work in sociology and cultural studies, and some natural scientists persuasively say in public there is shoddy work in their parts of science. *There is no excuse for shoddiness wherever it is found.* But we should at the same time cut each other a little bit of slack. To err is human, but it is as likely that we err in appreciating each others' intentions as it is that major blunders have been committed or that disciplinary hostility is at work. Before pointing fingers in the press or on public platforms, we might try conversations in a café or a pub. The likely result would be lower blood pressure and a less poisonous public culture.

Finally, as I suggested a while ago, scientists' metascientific statements often function in the specific context of *doing* science, of criticizing or applauding certain scientific claims or programs or disciplines. That is to say, they may not be pure expressions of institutional intentions to describe and interpret science but tools in saying what *ought* to be believed or done within science as a whole or within a particular discipline or subdiscipline. Viewed in that way, such statements not only can be taken seriously by students of science, they *must* be taken seriously, but *in a different way*—as part of the *topic* that the sociologist or historian means to describe and interpret.

The major conclusions I want to come to concern both the variability of scientists' metascientific statements and the nature of their relationship to what might loosely be called "science itself." Here I'd like to say—and again I can call on the additional authority of Einstein and Planck to say it—that the relationships between metascientific claims and the range of concrete scientific beliefs and practices are always going to be intensely problematic. "In the temple of science," Einstein said, "are many mansions" (1954, 224).[12] It is a modernist legacy, inherited from the methodological Public Relations Officers of the seventeenth century, that science is one, and, accordingly, that its "essence" can be captured by any one coherent and systematic metascientific statement, methodological or conceptual.[13] But, while the vision of scientific unification remains compelling to some, no plan for unification, and no account of the essence of science, carries conviction for more than a fraction of scientists. And that is one of my points.

So what happens if we follow the sentiments of many scientists (and incidentally that of increasing numbers of philosophers) that the sciences are many and diverse and that no coherent and systematic talk about a distinctive essence of science can make sense of the diversity or the concreteness of practices and beliefs? One thing that may happen is that we take a different view of the variability of metascientific statements, taken, that is, as statements about the distinctive nature of something called "science." We may want to say that different kinds of metascientific statements may pick out aspects of different kinds, or stages, or circumstances of the practices we happen to call scientific. Or different metascientific statements may contingently belong to the practices they purport to be about: as ideals, or norms, or strategic gestures signaling possible or desirable alliances. They may be true, or accurate, about science, but not globally true about science, just because no coherent and systematic statement could be globally true or accurate about science and could at the same time distinguish science from other forms of culture. Why ever should we expect that metascientific statements of any sort could hold for particle physics (which kind?) *and* for seismology *and* for the study of the reproductive physiology of marine worms? Some metascientific statements *might be* true about a range of scientific practices localized in time, place, and cultural context, but that is for us to find out, not to assume.

Something else follows from the recognition of diversity for current

12. For increasing pluralist sensibilities about science among philosophers, see, for example, Dupré 1993.

13. On this, see the classic essay by Isaiah Berlin 1998.

concern with antiscience. Because scientists' metascientific statements are diverse, and because it is possible that each picks out some real local features of some sciences, when considered from a certain point of view, the relationship between metascience and science is certainly problematic and at most contingent. For that reason alone, one can be allowed to dispute metascientific narratives of any kind without being understood to oppose science. If science is really as distinct from philosophy as some Defenders of Science insist it is, then it is puzzling in the extreme why they should be so upset when their favorite philosophy is criticized.[14] Natural science justifiably possesses enormous cultural authority; philosophy of science possesses rather little. Some tactical mistake is surely being committed when the Defense of Science appears as a celebration of a particular philosophy, still more when it celebrates versions that have been tried and long abandoned as faulty by philosophers themselves.

How to be antiscientific, then? I can now tell you some ways in which you cannot be coherently and effectively antiscientific. You cannot be against science because you dislike its supposedly unique, unifying, and universally effective Method. You cannot be against science because it is essentially materialistic or essentially reductionist. You cannot be against science because it is essentially "instrumental rationality" or, indeed, because it contains irrationality. You cannot be against science because it is a realist enterprise or because it is a phenomenological enterprise. You cannot be against it because it violates common sense or because it is a form of common sense. Nor can you be against it because it is essentially hegemonic, or essentially bourgeois, or essentially masculinist. And, of course, it should go without saying that you cannot be coherently *for* science for any of these reasons either.

A thought experiment, then a qualification, and finally some remarks on a sense in which one *can* be antiscientific in real, substantial, and constructive ways. First, the thought experiment. I, and some of my colleagues in the history and sociology of science, are methodological relativists. That is to say, I maintain, on the basis of empirical and theoretical work, that the standards by which different groups of practitioners assess

14. For example, Steven Weinberg's judgment that much philosophy of science "has nothing to do with science": "The fact that we scientists [which ones, please?] do not know how to state in a way that philosophers [which ones, please?] would approve what it is that we are doing in searching for scientific explanations does not mean that we are not doing something worthwhile. We could use help from professional philosophers in understanding what it is that we are doing, but with or without their help we shall keep at it" (1992, 167; also 29). (I am delighted to hear that. I would be very disturbed, indeed, if I thought that natural scientists were taking marching orders from philosophers!)

knowledge-claims are relative to context and that the appropriate methods to use in studying science should take that relativity into account. So far as the Scientific Method goes, like Peter Medawar and many other scientists, I am a skeptic. Further, this work leads me to believe that the natural world is probably extremely complex and that different cultures can stably and coherently classify and construe it in very different ways, according to their purposes and in light of the cultural legacies they bring to their engagements with the natural world. This position has been identified as antiscientific—motivated by ignorance and hostility—and, it is said, that people having such small faith in science should follow its logical conclusions: they should jump in front of cars or consult witch doctors rather than neurologists when their heads ache.

It is a silly and misguided argument, but nonetheless an interesting one to consider. I do not jump in front of cars and I do consult physicians when I need to do so. What does this prove? Not that I am insincere in my methodological relativism, or that I have contradicted myself, but that my genuine confidence in a range of modern scientific and technical practices and claims proceeds from different sources than my belief in some set of methodological metascientific stories. My confidence in science is very great: that is just to say that I am a typical member of the overall overeducated culture, a culture in which confidence in science is a mark of normalcy and which produces that confidence as we become and continue to be normal members of it.

I have been to the same sorts of schools as Alan Sokal, Steven Weinberg, Paul Gross, and Norman Levitt; we share other important cultural legacies and sensibilities; we probably vote the same way and like the same sorts of movies, though that's just a guess. Apart from our different academic disciplines, our institutional environment is much of a muchness; and if we met each other at a party with our name tags off, there's a decent chance that we'd hit it off pretty well. But, for all that, my professional confidence in a range of metascientific global stories about the Scientific Method, and its warrant for scientific effectiveness, is very low. So *this* is what is proved by my preference for physicians over witch doctors, for astronomers over astrologers: the grounds of my confidence in science have very little to do with metascientific stories, of any kind. And, arguably, the same situation obtains over a broad range of educated, and perhaps of not-so-educated people.

Now the qualification: in my academic work I have made, and I continue to make, claims about science that have an apparently global character, though to be honest I've become a bit more circumspect about making them as time has gone on. And I want to defend their character, pertinence, and legitimacy. So, for example, I've been known to say that the social dimension of science is constitutive and that trust is a necessary condition for the making and maintenance of scientific knowledge. These *are* metascientific statements, and they *are* meant to apply to all scientific practices that I know of. So am I not hoist on my petard? I don't think so. The reason is that when I say such things about science I am theorizing about the conditions for having knowledge of any kind; I am, so to speak, doing cognitive science without a license. What I am *not* doing is picking out a unique essence of science, meant to hold good for invertebrate zoology and for seismology and for particle physics (all kinds) and not to hold good for phrenology or accountancy or for the empirical and theoretical projects of everyday life. I may be right or wrong in the domain of theorizing-about-knowledge-of-any-kind, but I am not theorizing about a unique scientific essence. And that is the matter at issue.

Again the question: how to be antiscientific? As I said, being against the essence of science and being against one or other metascientific story uniquely about science are not very good ways of being antiscientific, nor do I find that my skepticism about the Scientific Method frees me in any way and to any degree from belief in the existence of electrons or in DNA as the biochemical basis of heredity. Those who are against the methodological or conceptual essence of science are against nothing very much in particular. And those who might be genuinely hostile to what they take to be the essence of science are probably just as ineffective as they are misguided. Who reads this stuff anyway? In order to corrupt the youth of Athens, you first have to get this stuff in their hands, then you have to get them to read it, and understand it, and care about it; then you have to persuade them—against the background of everything else they've been told—that you're right. Not such an easy business, really, as any teacher in my line of work knows.

But being against something in particular about science is both possible and legitimate. How to be against something in particular about science *should one wish to be so*? Here again it is good to listen to what some scientists themselves have to say. And if we listen to scientists (other than those who are taking the lead in the science wars), what we can hear is not a global defense of science, nor, of course, a global criticism of science. Rather, we can hear local criticisms of certain tendencies *within* science, or within parts of it—criticisms that are often substantial and vehemently expressed.

Some scientists are now violently critical of what *they* take to be the shallowness of reductionist programs, the tyrannizing and stultifying effects of bureaucratization in science, the dedicated following of scientific fashion and the attendant loss of the Big Picture and of imagination, the hegemony of Big Science at the expense of Little, the incompetence of the peer review system, the commercialization of science and the attendant ethical and intellectual erosion, and many other ills *they* diagnose in the contemporary Body Scientific. Some of these internal criticisms happen to look to professional metascience and even to the history of science for aids in understanding how current arrangements came to be and as tools in making things better; many do not.

It is not difficult to find these public internal criticisms: recent issues of biological periodicals are full of them, and memoirs and reflections by eminent scientists—including those by E. O. Wilson, Erwin Chargaff, Gunther Stent, and Richard Lewontin—are another rich seam of such things. The striking thing, given the ultimate vacuity of the science wars, is just how little professional metascientists have concerned themselves with these internal contests, and, indeed, how little sociologists and historians have even noticed them as topics. That is almost certainly a Bad Thing: if being against science is, as I am suggesting, being against nothing very much in particular, being against the current peer review system, or against the hegemony of Big Science, or against the way in which clinical trials are constituted and funded, is being against something substantial and important. Is it sociologists' and historians' role to take sides in such debates? I don't think so (though I know of some sociologists who disagree). But these debates do offer a venue in which we can have interesting and substantial conversations with our scientist-colleagues. It would be mutually beneficial to have these conversations.

Finally, we need to remember that professional metascientists, like professional scientists, are also citizens. We are equal members—many of us—of institutions of higher education, and we all pay our share in the state support of scientific research. So far as the first type of citizenship is concerned, no one, I think, should rule it out of order, or identify it as lèse majesté, to take one side or another in university discussions over, for example, how much science should be taught in the curriculum or how scientific subjects should be taught. If one wants to say (as I do *not*) that there's too much science in the required curriculum, or if one wants to say (as I *do*) that the philosophical, historical, or sociological dimensions should have a place in the science curriculum, then one should be free to do so. And, should one want to make such arguments, one should not have to face accusations of being antiscientific.

Similarly, as citizens paying the bill for much scientific research, one should be free to say if one wants—and on an informed basis—that the

Superconducting Supercollider cost too much relative to its advertised benefits, or that too much money is going to a cure for AIDS and too little for an AIDS vaccine, or that governments have got their priorities wrong as between AIDS research and diarrhea research, or that some science supported by the public treasure is trivial or intellectually unimaginative, or that the links between publicly funded science and the commercial world are becoming worrying. And one should be able to say such things—again, if one wants—without being denounced as antiscientific. Some scientists say such things on a professional basis, and some citizens may want to say such things as responsible members of democratic societies. They must be free to do so, not intimidated into deferential silence.

My fear is that, if we carry on in our present courses, the ultimate and consequential casualties of the science wars will not be the job security of sociologists of science, but free, open, and informed public debate about the health of modern science. And the health of science ultimately depends on that debate.

Here are the sources for the notorious metascientific claims at the beginning of this essay:

1. Many sources, including Peter B. Medawar (immunologist), *The Art of the Soluble* (London: Methuen, 1957), 132; James B. Conant (chemist), *Science and Common Sense* (New Haven: Yale University Press, 1951), 45; Lewis Wolpert (biologist), *A Passion for Science* (Oxford: Oxford University Press, 1988; compiled with Alison Richards), 3; Richard Lewontin, "Billions and Billions of Demons," *New York Review of Books* 44, no. 1 (9 January 1997):28–32 (on page 29: "The case for the scientific method should itself be 'scientific' and not merely rhetorical").

2. Erwin Chargaff (biochemist), *Heraclitean Fire: Sketches from a Life before Nature* (New York: Rockefeller University Press, 1978), 138.

3. Edward O. Wilson (entomologist, sociobiologist), *Naturalist* (New York: Warner Books, 1995), 210.

4. Niels Bohr, quoted in Abraham Pais, *Niels Bohr's Times, in Physics, Philosophy, and Polity* (Oxford: Clarendon Press, 1991), 314.

5. Albert Einstein (physicist), *Out of My Later Years* (New York: Philosophical Library, 1950), 96; also Einstein, *Ideas and Opinions* (New York: Crown Publishers, 1954), 355. I have here slightly paraphrased Einstein's original statement that the bases of physics cannot be inductively secured from experience, but "can only be attained by free invention." Geometrical axioms—the bases of the deductive structure of physics—are, Einstein said, "free creations of the human mind" (1954, 234).

6. Jacob Bronowski (mathematician), "Science is Human," in *The Humanist Frame,* ed. Julian Huxley (New York: Harper and Brothers, 1961), 83–94 (quote, page 88). I have here altered the first-person "we" to the third-person "scientists."

7. Warren Weaver (mathematician and scientific administrator), "Science and People," in Paul C. Obler and Herman A. Estrin, eds., *The New Scientist: Essays on the Methods and Values of Modern Science* (Garden City, NY: Anchor, 1962), 95–111 (quote, page 104).

8. Gunther Stent (biochemist), interviewed in Lewis Wolpert and Alison Richards, *A Passion for Science* (Oxford: Oxford University Press, 1988), 116.

9. Brian Petley (physicist), *The Fundamental Physical Constants* (Bristol: Adam Hilger, 1985), 2: "Modern physics is based on some intrinsic acts of faith, many of which are embodied in the fundamental constants."

10. Richard Lewontin (evolutionary geneticist), "Billions and Billions of Demons" (see reference 1), 31: "[The public] take the side of science *in spite* of the patent absurdity of some of its constructs, *in spite* of its failure to fulfill many of its extravagant promises of health and life, *in spite* of the tolerance of the scientific community for unsubstantiated just-so stories, because we have a prior commitment, a commitment to materialism."

11. Richard Lewontin, Steven Rose (neurobiologist), and Leon J. Kamin (psychologist), *Not in Our Genes: Biology, Ideology, and Human Nature* (New York: Pantheon, 1984), 33; see also "[T]he internalist, positivist tradition of the autonomy of scientific knowledge is itself part of the general objectification of social relations that accompanied the transition from feudal to modern capitalist societies" (33). It would not be easy to find such a sweepingly didactic statement of this kind expressed by present-day historians or sociologists of science!

Chapter 9

PHYSICS AND HISTORY

Steven Weinberg

I am going to discuss the uses that historical and scientific knowledge have for each other, but first I want to take up what may be a more unusual topic: the dangers that history poses for physics, and physics for history.

The danger in history for the work of physics is that, in contemplating the great work of the past—great heroic revolutions like relativity, quantum mechanics, and so on—we develop such respect for them that we become unable to reassess their place in a final physical theory. General relativity provides a good example.

As developed by Einstein in 1915, general relativity appears almost logically inevitable. One of its fundamental principles, the equivalence of gravitation and inertia, says that there is no difference between gravity and effects of inertia such as centrifugal force. This principle of equivalence can be reformulated as the principle that gravity is just an effect of the curvature of space and time—a beautiful principle from which Einstein's theory of gravitation follows almost uniquely. But there is an "almost" here. To arrive at the equations of general relativity, Einstein in 1915 had to make an additional assumption; he assumed that the equations of general relativity would be of a particular type. He assumed

This chapter consists of material from two previously published works ("Physics and History," reprinted by permission of *Daedalus,* Journal of the American Academy of Arts and Sciences, from the issue entitled, "Science in Culture," winter 1998, vol. 127, no. 1, 151–64; and "The Revolution That Didn't Happen," *New York Review of Books,* 8 October 1998, 48–52), which was reorganized by the editors of the current book and finally revised by the author.

that they would be second-order partial differential equations, not higher order equations. They would not involve, for instance, third-order rates, which give the rates of change of the rates at which the rates are changing. This may seem like a technicality. It is certainly not a grand principle like the principle of equivalence; it is just a limit on the sorts of equations that will be allowed in the theory. Now why did Einstein make this assumption—this very technical assumption, with no philosophical underpinnings? Well, that is what people were used to at the time: the equations of Maxwell that govern electromagnetic fields and the wave equations that govern the propagation of sound are all second-order differential equations. For a physicist in 1915, it was a natural assumption. If a theorist does not know what else to do, it is a good tactic to assume the simplest possibility because this is more likely to give a theory that one could actually solve and thereafter decide whether or not it agrees with experiment. In Einstein's case, this tactic worked.

But this kind of pragmatic success does not in itself provide a rationale for the assumption, at least not one that would satisfy Einstein, of all people. Einstein's goal was never simply to find theories that fit the data. Remember, it was Einstein who said that the purpose of the kind of physics that he did was "not only to know how nature is and how her transactions are carried through, but also to reach as far as possible the utopian and seemingly arrogant aim of knowing why nature is thus and not otherwise" (1929, 126). He certainly was not achieving that goal when he arbitrarily assumed that the equations for general relativity were second-order differential equations. He could have made them fourth-order differential equations, but he didn't.

Our current perspective, which has been developing gradually over the last fifteen or twenty years, differs from that of Einstein. Many of us regard general relativity as nothing but an effective field theory—an approximation to a more fundamental theory that is valid in the limit of large distances, probably including any distances that are larger than the scale of an atomic nucleus. Indeed, if one supposes that there really are terms in the Einstein equations that involve rates of fourth or higher order, such terms would still play no significant role at sufficiently large distances. This is why Einstein's tactic worked. There *is* a rational reason for assuming the equations are second-order differential equations, which is that any terms in the equations involving higher order rates would not make much of a difference in any astronomical observations, but as far as I know this was not Einstein's rationale.

This may seem like rather a minor point to raise here, but in fact the most interesting work today in the study of gravitation is precisely in contexts where the presence of higher order rates in the field equations *would* make a big difference. The most important problem in the quantum theory of gravity arises from the fact that when you do various calculations, for instance of the probability that a gravitational wave will be scattered by another gravitational wave, you get answers that turn out to be infinite. Another problem in the classical theory of gravitation arises from the presence of singularities: matter can apparently collapse to a point in space with infinite energy density and infinite space-time curvature. These absurdities, which have been exercising the attention of physicists for many decades, are precisely problems that involve gravity at very short distances—not the large distances of astronomy, but distances much smaller than the size of an atomic nucleus.

Now, from the point of view of modern effective field theory, there are no infinities in the quantum theory of gravity. The infinities are canceled in exactly the same way that they are in all our other theories, by being just absorbed into a redefinition of parameters in the field equations. But this works only if we include terms involving rates of fourth order and all higher orders. The old problems of infinities and singularities in the theory of gravitation cannot be dealt with by taking Einstein's theory seriously as a fundamental theory. From the modern point of view—or, if you like, from my point of view—Einstein's theory is nothing but an approximation valid at long distances, which cannot be expected to deal successfully with infinities and singularities. Yet professional quantum gravitationalists (if that is the word) go on spending their whole careers studying the applications of the original Einstein theory, the one that only involves second-order differential equations, to problems involving infinities and singularities. Elaborate formalisms have been developed that aim to look at Einstein's theory in a more sophisticated way, in hopes of somehow making the infinities or singularities go away. I think that this ill-placed loyalty to general relativity in its original form persists because of the enormous prestige that the theory earned from its historic successes.

We have to be wary lest the great heroic ideas of the past weigh upon us and prevent us from seeing things in a fresh light, and it is just those ideas that were most successful of which we should be most wary. I could give other examples of this sort. For instance, there is an approach to quantum field theory called second quantization which fortunately no longer plays a significant role in our research but docs go on playing a role in the way that textbooks are written. Second quantization goes back to an idea in a 1927 paper by Jordan and Klein, the idea that after one has introduced a wave function in quantizing a theory of particles, one should then quantize the wave function. Surprisingly, many people think that this is the way to look at quantum field theory, but it isn't.

We have to expect the same fate for our present theories. The standard model of weak, electromagnetic, and strong forces that describes nature under conditions that can be explored in today's accelerators may itself be different in the future—it will not disappear or be proved wrong, but it may be looked at in quite a different way. Most particle physicists now think of the standard model as only an effective field theory that provides a low-energy approximation to a more fundamental theory.

Enough about the danger of history to science. Now I would like to say something about the danger of scientific knowledge to history. The danger arises from a tendency to imagine that discoveries are made according to the logic of our present understanding. By assuming that scientists of the past thought about things the way we do, we not only make mistakes, but we also lose appreciation for the difficulties, for the intellectual challenges, that they faced. One has to look at things as they really were in their own time.

This, of course, also applies to political history. There is a term "Whig interpretation of history," which was invented by Herbert Butterfield in a lecture in 1931. As Butterfield explained it, "the Whig historian seems to believe that there is an unfolding logic in history." He went on to attack the person he regarded as the archetypal Whig historian, Lord Acton, for his view of history as a means of passing moral judgments on the past. Acton wanted history to serve as the "arbiter of controversy, the upholder of that moral standard which the powers of earth and religion itself tend constantly to depress. . . . It is the office of historical science to maintain morality as the sole impartial criterion of men and things." Butterfield went on to say that "[i]f history can do anything it is to remind of us of those complications that undermine our certainties, and to show us that all our judgments are merely relative to time and circumstance. . . . [W]e can never assert that history has proved any man right in the long run. We can never say that the ultimate issue, the succeeding course of events, or the lapse of time have proved that Luther was right against the Pope or that Pitt was wrong against Charles James Fox" (Butterfield 1951, 75).

This is the point where the historian of science and the historian of politics should part company. The passage of time has shown that, for example, Darwin was right against Lamarck, the atomists were right against Ernst Mach, and Einstein was right against the experimentalist Walter Kaufmann, who presented data contradicting special relativity. To put it another way, Butterfield was correct: there is no sense in which

Whig morality (much less the Whig party) existed at the time of Luther. In contrast, it is true that natural selection was working during the time of Lamarck, and the atom did exist in the days of Mach, and fast electrons behaved according to the laws of relativity even before Einstein. Present scientific knowledge has the potentiality of being relevant in the history of science in a way that present moral and political judgments may not be relevant in political or social history [16].

Many historians, sociologists, and philosophers of science have taken the desire for historicism, the worry about falling into a Whig interpretation of history, to extremes. To quote Holton, "Much of the recent philosophical literature claims that science merely staggers from one fashion, conversion, revolution, or incommensurable exemplar to the next in a kind of perpetual, senseless Brownian motion, without discernible direction or goal" (1996, 22). I made a similar comment in a talk to the American Academy of Arts and Sciences, noting in passing that there are people who see scientific theories as nothing but social constructions. A transcript of the talk fell into the hands of someone who, over twenty years ago, had been closely associated with a development known as the sociology of scientific knowledge, or SSK. He wrote me a long unhappy letter in which, among other things, he complained about my remark that the Strong Programme initiated at the University of Edinburgh embodied a really radical social constructivist view in which scientific theories are nothing but social constructions. He sent me a big pile of essays and said that if I read them, I would see that he and his colleagues do recognize that reality plays a role in our world. I took this criticism to heart and read the essays; I also looked back over some old correspondence that I had had with Harry Collins, who for many years led the well-known SSK group at the University of Bath. I tried to look at these materials from as sympathetic a point of view as I could, to try to understand what they were saying and to assume that they must be saying something that is not absurd.

In fact I found a point of view that is not absurd on its face described (though not espoused) in an article by David Bloor, who is one of the Edinburgh group, and also in my correspondence with Harry Collins. As I understand it, there is a position called "methodological idealism" or "methodological antirealism," which holds that historians or sociologists should take no position on what is ultimately true or real. Instead, they should simply take it as a guiding principle in their work, not to use today's scientific knowledge, but rather to try to look at nature as it must have been viewed by the scientists under study at the time that they were working. In itself, this is not an absurd position. In particular, it can guard us against the kind of silliness that, for instance, I recall having been guilty of, in the days when I assumed Kepler had derived his laws of planetary motion directly from observations of planetary motions in the sky.

But the attitude of methodological antirealism bothered me, though for a while I couldn't point to what I found wrong with it. In preparing this essay I have tried to think this through, and I have come to the conclusion that there are a number of minor things wrong with methodological antirealism: it can cripple historical research, it is often boring, and it is basically impossible. It also has a major drawback: in an almost literal sense, it misses the point of the history of science.

Let me first cover the minor points. If it were possible really to reconstruct everything that happened during some past scientific discovery, then it might be helpful to forget everything that has happened since, but in fact much of what happened will always be unknown to us. For example, J. J. Thomson, in the experiments that made him known as the discoverer of the electron, was measuring a certain crucial quantity, the ratio of the electron's mass to charge. As always happens, he found a range of values. Although he quoted various values in his published work, the values he would always refer to as his favorite results were those at the high end of the range. Why did Thomson quote the high values as his favorite values? It is possible that Thomson knew that on those days he was more careful, or hadn't bumped into the laboratory table, or had had a good night's sleep the night before. But there is another possibility: that his first values were at the high end of the range, and he was determined to show that he had been right at the beginning. Now which explanation is correct? There is simply no way of reconstructing the past. Not his notebooks, not his biography, nothing will allow us now to reconstitute those days in the Cavendish Laboratory and find out on which days he was more clumsy or sleepy than usual. There is one thing that we do know, however: the actual value of the ratio of the electron's mass to charge. It was the same in Thomson's time as it is in our own. We know that this actual value is *not* at the high end of the range of Thomson's experimental values, but at the low end, which strongly suggests that the measurements that gave high values were not in fact more careful and that therefore it is more likely that Thomson quoted these values because he was trying to justify his first measurements [15, 21].

This is a trivial example of the use of present scientific knowledge in the history of science, because here we are just talking about a number, not a natural law or an ontological principle. I chose this example because it shows so clearly that to decide to ignore present scientific knowledge is often to throw away a valuable historical tool.

A second minor drawback of methodological antirealism is that a reader who does not know anything about our present understanding of nature is likely to find the history of science terribly boring. For instance, a historian might describe how in 1911 the Dutch physicist Kamerlingh Onnes was measuring the electrical resistance of a sample of cold mercury and thought that he had found a short circuit. The historian could go on for pages and pages describing how Onnes searched for the short circuit and how he took apart the wiring and put it back together again but couldn't find the source of the short circuit. Could anything be more boring than to read this description if one did not know in advance that there *was* no short circuit? That what Onnes was observing was in fact the vanishing of the resistance of mercury when cooled to a certain temperature and that this was nothing less than the discovery of superconductivity? Of course, it is impossible today for a physicist or a historian of physics not to know about superconductivity; we are quite incapable while reading about the experiments of Kamerlingh Onnes to imagine that his problem was nothing but a short circuit. But even if one had never heard of superconductivity, the reader would know that there was something going on besides a short circuit—why else would the historian bother with these experiments? Plenty of experimental physicists have found short circuits, and no one studies them [15, 21]!

But these are minor issues, and I don't want to dwell on them. I think the main drawback of methodological antirealism is that it misses the point about the history of science that makes it different from other kinds of history: Even though a scientific theory is in a sense a social consensus, it is unlike any other sort of consensus in that it is culture-free and permanent [15, 20].

This is just what many sociologists of science deny. David Bloor said in a talk at Berkeley that "the important thing is that reality underdetermines the scientists' understanding." I gather he means that although he recognizes that reality has some effect on what scientists do—scientific theories are not "nothing but" social constructions—scientific theories are also not what they are simply because that is the way nature is. In a similar spirit, Stanley Fish argued that the laws of physics are like the rules of baseball (1996). It is true that both are certainly conditioned by external reality—after all, if baseballs moved differently under the influence of Earth's gravity, the rules would call for the bases to be closer together or farther apart. However, the rules of baseball also reflect the way that the game developed historically and the preferences of players and fans.

Now, what Bloor and Fish say about the laws of nature *does* apply while these laws are being discovered. As Holton has shown in work on Einstein, Kepler, and superconductivity, many cultural and psychological influences enter in scientific work. But the laws of nature are not like the rules of baseball. They are culture-free and they are permanent—not as they are being developed, not as they were in the mind of the scientist who first discovers them, not in the course of what Latour and Woolgar call "negotiations" over what theory is going to be accepted—but in their final form, in which cultural influences are refined away [22, 32]. I will even use the dangerous words "nothing but": aside from inessentials like the mathematical notation we use, the laws of physics as we understand them now are nothing but a description of reality.

I can't *prove* that the laws of physics in their mature form are culture-free. Physicists today live embedded in the Western culture of the late twentieth century, and it is natural to be skeptical if I say that our understanding of Maxwell's equations or quantum mechanics or relativity or the standard model of elementary particles is culture-free. However, I am convinced that it is so, because the purely scientific arguments for these theories seem to me overwhelmingly convincing. I can add that as the typical background of physicists has changed, in particular as the number of women and East Asians in physics has increased, the nature of our understanding of physics has not changed. These laws in their mature form have a toughness that resists cultural influence.

The other thing about the history of science that is different from political or artistic history (and that reinforces my remarks about the influence of culture) is that the achievements of science become permanent. Now that may seem to contradict what I said at the beginning of this chapter: that we now look at general relativity in a different way than Einstein did, and that even now we are beginning to look at the standard model differently than we did when it was being developed. But what changes is our understanding of why the theories are true and also our understanding of their scope of validity. For instance, at one time we thought that there was an exact symmetry in nature between left and right, but then it was discovered that this is only true in certain contexts and to a certain degree of approximation. But the theory of a symmetry between right and left was not a simple mistake, it has not been abandoned, we just understand it better. Within its scope of validity, this symmetry has become a permanent part of science, and I can't see that this will ever change.

Finally, let me turn to the use of history in physics. History plays a special role for elementary particle physicists like myself. Our vision of history is quite different from that of those working in most of the other sciences. Other scientists look forward to an endless future of finding interesting problems—understanding consciousness or turbulence or high temperature superconductivity—that will go on forever. In contrast, many elementary particle physicists think that our work in finding deeper and deeper explanations will come to an end in a final theory toward which we are working. Our aim is to put ourselves out of business. This gives a historical dimension to our choice of the sort of work on which to concentrate. We tend to seek problems that further this historic goal, not just work that is interesting or useful or that influences other fields but work that is historically progressive, that moves us toward the goal of a final theory.

Of course, the entire concept of science as progress has been strongly questioned, especially since the appearance of Thomas Kuhn's famous book *The Structure of Scientific Revolutions*. In it, Kuhn argued that in scientific revolutions it is not only our scientific theories that change, but the very standards by which scientific theories are judged, so that the paradigms that govern successive periods of normal science are incommensurable. He went on to reason that, since a paradigm shift means complete abandonment of an earlier paradigm, and there is no common standard to judge scientific theories developed under different paradigms, there can be no sense in which theories developed after a scientific revolution can be said to add cumulatively to what was known before the revolution. Only within the context of a paradigm can we speak of one theory being true or false. He concluded, tentatively, "We may, to be more precise, have to relinquish the notion explicit or implicit that changes of paradigm carry scientists and those who learn from them closer and closer to the truth" (170). More recently, in his Rothschild Lecture at Harvard in 1992, Kuhn remarked that it is hard to imagine what can be meant by the phrase "closer to the truth."

All this is wormwood to scientists like myself, who think the task of science is to bring us closer and closer to objective truth. But Kuhn's conclusions are delicious to those who take a more skeptical view of the pretensions of science. If scientific theories can only be judged within the context of a particular paradigm, then in this respect the scientific theories of any one paradigm are not privileged over other ways of looking at the world, such as shamanism or astrology or creationism. If the transition from one paradigm to another cannot be judged by any external standard, then perhaps it is culture rather than nature that dictates the content of scientific theories.

Kuhn did not deny that there is progress in science, but he denied that it is progress *toward* anything. He often used the metaphor of biological evolution: scientific progress for him was like evolution as described by Darwin, a process driven from behind, rather than pulled toward some fixed goal to which it grows ever closer. For him, the natural selection of scientific theories is driven by problem solving. When, during a period of normal science, it turns out that some problems can't be solved using existing theories, then new ideas proliferate, and the ideas that survive are those that do best at solving these problems. But according to Kuhn, just as there was nothing inevitable about mammals appearing in the Cretaceous period and outsurviving the dinosaurs when a comet hit the Earth, so also there's nothing built into nature that made it inevitable that our science would evolve in the direction of Maxwell's equations or general relativity. Kuhn recognized that Maxwell's and Einstein's theories are better than those that preceded them, in the same way that mammals turned out to be better than dinosaurs at surviving the effects of comet impacts; but when new problems arise they will be replaced by new theories that are better at solving those problems, and so on, with no overall improvement.

To defend this conclusion, Kuhn argued that all past beliefs about nature have turned out to be false and that there is no reason to suppose that we are doing better now. Of course, he knew very well that physicists today go on using the Newtonian theory of gravitation and motion and the Maxwellian theory of electricity and magnetism as good approximations that can be deduced from more accurate theories—we certainly don't regard Newtonian and Maxwellian theories as simply false in the way that Aristotle's theory of motion or the theory that fire is an element ("phlogiston") are false. Kuhn himself, in his earlier book on the Copernican revolution, told how parts of scientific theories survive in the more successful theories that supplant them and seemed to have no trouble with the idea. Confronting this contradiction, Kuhn in *Structure* gave what for him was a remarkably weak defense: that Newtonian mechanics and Maxwellian electrodynamics as we use them today are not the same theories they were before the advent of relativity and quantum mechanics because then they were not known to be approximate and now we know that they are. It is like saying that the steak you eat is not the one that you bought because now you know it is stringy and before you didn't.

It is important to keep straight what does and what does not change in scientific revolutions, a distinction that is not made in *Structure*. There is a "hard" part of modern physical theories ("hard" meaning "durable," like bones in paleontology or potsherds in archeology) that usually consists of the equations themselves, together with some understandings about what the symbols mean operationally and about the sorts of phenomena to which they apply. Then there is a "soft" part; it is the vision of reality that we use to explain to ourselves why the equations work. The soft part does change; we no longer believe in Maxwell's ether, and we know that there is more to nature than Newton's particles and forces. The changes in the soft part of scientific theories also produces changes in our understanding of the conditions under which the hard part is a good approximation. But after our theories reach their mature forms, their hard parts represent permanent accomplishments. If you have bought one of those tee shirts with Maxwell's equations on the front, you may have to worry about it going out of style, but not about it becoming false. We will go on teaching Maxwellian electrodynamics as long as there are scientists. I can't see any sense in which the increase in scope and accuracy of the hard parts of our theories is not a cumulative approach to truth.

Kuhn's view of scientific progress would leave us with a mystery: why does anyone bother? If one scientific theory is only better than another in its ability to solve the problems that happen to be on our minds today, then why not save ourselves a lot of trouble by putting these problems out of our minds? We don't study elementary particles because they are intrinsically interesting, like people. They are not—if you have seen one electron, you've seen them all. What drives us onward in the work of science is precisely the sense that there are truths out there to be discovered, truths that once discovered will form a permanent part of human knowledge.

When I said earlier that the social constructivists have missed the point, I had in mind an image of what is known in mathematical physics as the approach to a fixed point. There are various problems in physics that deal with motion in some sort of space, governed by equations that dictate that wherever you start in the space, you always wind up at the same point, known as a fixed point. Ancient geographers had something similar in mind when they said that all roads lead to Rome. I think that physical theories are like fixed points, toward which we are attracted. Our starting points may be culturally determined, our paths may be affected by our personal philosophy, but the fixed point is there nonetheless. It is something toward which any physical theory moves, and when we get there we know it, and then we stop.

The kind of physics that I've done for most of my life, while I have been working in the theory of fields and elementary particles, is moving toward a fixed point. But this fixed point is unlike any other in science. That final theory toward which we are moving will be a theory of unrestricted validity, a theory which is applicable to all phenomena throughout the universe, a theory which, when we discover it, will be a permanent part of our knowledge of the world. Then our work as elementary particle physicists will be done, and will become nothing but history.

Chapter 10

SCIENCE STUDIES AS EPISTEMOGRAPHY

Peter Dear

EPISTEMOGRAPHICAL ENDEAVORS IN SCIENCE STUDIES

The famous remark made by Molière's character in *Le bourgeois gentilhomme* comes forcibly to mind: "Good heavens! For more than forty years I have been speaking prose without knowing it."[1] This essay suggests that, in the case of science studies, its practitioners have all along been engaged in an enterprise that we, too, have not known how to name, and that it is "epistemography." Unfortunately, this lapse has compounded the misunderstandings between science studies and its would-be critics, who have themselves often failed to recognize what it is that they are attacking.

The ambiguities of the catch-all label "science studies" are not accidental: such crypto-disciplinary names tend to be hobbled from the start by the need to draw in as many constituencies as possible, thereby running the risk of yoking together quite distinct intellectual endeavors. If (no doubt, contestably) we date science studies from the later 1970s, we immediately perceive a genealogy. Before science studies, there was a fairly well-established specialty called history and philosophy of science. There was also a subfield of sociology known as sociology of science. Science studies has now swallowed up large amounts of what once fell into those earlier categories, without always thoroughly digesting them. But the parts that were left behind tell their own stories.

1. Act II, scene 4.

In his 1983 book *Wittgenstein: A Social Theory of Knowledge,* David Bloor refers to Wittgenstein's description of his own work as one of the "heirs to the subject which used to be called philosophy" (183). Thanks in part to the efforts of Bloor himself, an heir to the specific domain of the philosophy of science had by then become evident: *some* of it had become a part of science studies, specifically the approach known as the sociology of scientific knowledge (SSK); some of it had not. The parts that had not were certain *evaluative* and *normative* enterprises in the philosophy of science that strove to justify particular kinds of scientific procedure as epistemologically superior to others and hence more capable of providing reliable knowledge about the world. But the parts that had entered into science studies were the descriptive ambitions to codify and generalize, as much as possible, the ways in which scientific knowledge was actually made in various disciplines and in various times and places in history. Typically, such ambitions in the philosophy of science were yoked to attempts at using the descriptions as evidence for one or another normative theory of science, a so-called naturalistic approach (Donovan, Laudan, and Laudan 1988; Fine 1986); in science studies, by contrast, a self-consciously neutral stance has been the norm. The latter approach, often described as "relativist," has spawned much argument at cross-purposes between some science studies practitioners and some philosophers, but the form in which it appears most often in practice is that labeled by Harry Collins as "methodological relativism."[2]

Much of the most intense vitriol in the science wars has been spilled by critics of what is often pilloried as "postmodernist science studies," or something similar with the word "postmodern" in it.[3] Unfortunately, as Philip Kitcher notes, the people who are set up as the straw men of this enterprise are, as often as not, scholars that avowed practitioners of science studies would regard as marginal—and when they are not, they are advocates of positions given little countenance by the majority of their disciplinary peers. Kitcher pithily represents this as "the assumption that all contributors to science studies are really variations on Sandra Harding" (Kitcher 1998). This is where the lack of an adequate disciplinary label becomes a real liability. In the recent collection edited by Noretta Koertge, *A House Built on Sand: Exposing Postmodernist Myths about Science* (1998), "science studies," "cultural studies of science," as

2. Collins seems to have reached this position from a more aggressive earlier stance by the time of his 1985 *Changing Order* (Collins 1992).

3. See, for a notable recent example, Koertge 1998.

well as the title's generalized tilt at "postmodernism" are all terms bandied about vaguely, and largely interchangeably.[4] Such a practice is damaging not because there are no real targets deserving of criticism among the book's mélange of victims (there are quite a lot), but because the assault is indiscriminate and uncharitable. That is, the baby will get thrown out with the bath water if anything that can be caricatured as nonsensical is automatically rejected without a serious attempt having been made to understand it. It is akin to a summary rejection of physics simply because some of the conclusions of special relativity can appear counterintuitive.[5]

The fact is that in amongst the examples of sophisticated foolishness assailed in volumes such as Koertge's are to be found a number of serious scholarly studies that are given short shrift because their arguments are not understood. None of them is above criticism, which is the normal procedure in any academic discipline, and none of them has been spared the occasional wrath of sympathetic scholars. But there is an enormous difference between, on the one hand, a criticism based on taking an author's endeavor seriously and finding it faulty and, on the other hand, condemning an entire academic specialty as self-regarding nonsense while failing even to notice what problems an author is trying to address. As long as there is no better disciplinary label than "science studies" available, apples and oranges will continue to be crated together, wheat and chaff will remain inseparable, and standards of academic discourse will remain at a low level, both intellectually and morally. This essay, therefore, is a plea for the recognition of an already-existing scholarly enterprise, the prose that many of us have long been attempting to speak: it is an enterprise that is at its core *epistemographical.*

The term "epistemography" is intended to bring some clarity to the discussion by proposing a loose grouping of the most central and characteristic kinds of work currently encompassed by the label "science studies." The grouping strategy relies on making explicit the following recognition: the field of science studies is driven by attempts to understand what science, as a human activity, actually is and has been. Epistemography is the endeavor that attempts to investigate science "in the field," as it were, asking such questions as these: What counts as scientific knowledge? How is that knowledge made and certified? In what ways is it used or valued? "Epistemography" as a term signals that descriptive

4. The "cultural studies" label is applied by Sullivan (1998; see especially page 92).

5. Cf. the remarks by Michael Lynch (1997) on the rhetorical career of Alan Sokal's spoof in *Social Text.*

focus, much like "biography" or "geography."[6] It designates an enterprise centrally concerned with developing an empirical understanding of scientific knowledge, in contrast to *epistemology,* which is a prescriptive study of how knowledge can or should be made.

Such a move helps to clarify the vexed issue of relativism in science studies. If we translate this issue into a question of relativism in *epistemography,* the stakes become evident—both what they are and, crucially, what they are not. Epistemographical relativism is a species of methodological relativism. Relativism is an essential part of epistemography because it is employed not just for heuristic reasons, but also, specifically, for reasons of sheer relevance. A basic logical point lies at its foundation: people do not believe propositions to be true *because* those propositions are in fact true; instead, they believe things for various reasons that an epistemographer is interested to uncover [14]. If someone asks me why I believe that the Earth is round, it would little serve my case to reply that I believe it because it's true. I would likely adduce various empirical grounds of the kind employed by ancient Greek philosophers, to do with the disappearance of ships' hulls over the horizon or the change in apparent altitude of the pole star as the observer travels north or south; or else modern arguments to do with photographs made from outer space. Whatever the evidence and arguments might be, they would count as at least part of the explanation for my belief, regardless of their plausibility to other people.[7] Logically, the *truth* of the belief could never explain it, even if we were God and happened to *know* the absolute truth.

The complete analytical divorce between truth and belief means that the epistemic relativism attaching to beliefs has no bearing on any kind of position that maintains that truth itself is somehow "relative." An epistemographer could maintain a belief in absolute truth and still be a methodological relativist, because the relativism only applies to understanding what particular groups of people believe and why they believe it. That is, the divorce means that an epistemographer's commitment to absolute truth could go well beyond the view that there is absolute truth but that only God knows it (a position akin to Paul Feyerabend's "metaphysical realism" [1978]). A realist, but methodologically relativ-

6. The suffix "-ography" should not be taken to indicate anything more specific than "description" in the widest sense; it need not imply spatial description (akin to "cartography"), for example, although it could well do so in particular cases.

7. The qualification in this statement is due to the fact that beliefs are usually best approached through an understanding of their collective status: an individual's beliefs, and their associated criteria of credibility, would themselves be made sense of by reference to the linguistic and other social conventions by which the individual operates (and knows, and understands . . .).

ist, epistemographer can always, away from the job, condemn a belief as false and ill-founded with no self-contradiction at all:[8] the holders of the belief, the epistemographer says, are wrong, but due to their peculiar (and describable) ways of seeing things they can never, perhaps, be persuaded of their error.[9] The epistemographical work proceeds in the same way regardless of such metalevel views on relativism.

Nonetheless, even after recognizing epistemography as the disciplinary core of science studies, we can still anticipate fierce criticisms based not on philosophical or epistemological grounds, but on moral grounds. An examination of a recent bitter exchange in the *New York Review of Books* illustrates the point well precisely because of its extreme character, which borders on caricature.

PERUTZ VERSUS GEISON: EPISTEMOLOGY VERSUS EPISTEMOGRAPHY?

In December of 1995, Professor Max Perutz of Cambridge University published a review (1995) of Gerald L. Geison's widely praised book *The Private Science of Louis Pasteur* (1995).[10] Bearing the appropriate editorial heading "The Pioneer Defended," Perutz's review took issue with Geison's portrayal of Pasteur on grounds based chiefly in the text of Geison's book itself: that is, Perutz did not claim that Geison's factual account was in error as regards what was done and when; instead, taking all that as a given, Perutz objected to the evaluations of Pasteur's work found in, or else inferred from, Geison's writing. All of Perutz's counter-evaluations, like the evaluations that he attributed to Geison himself, turned on moral issues—including those evaluations regarding the scientific credentials of Pasteur's work.

In his article, Perutz introduces Pasteur as "the father of modern hygiene, public health, and much of modern medicine" and briefly summarizes the false or superstitious beliefs that Pasteur eradicated by his own discoveries. Quickly, however, the real issue is introduced: Pasteur,

8. See, for relevant and well-established discussion, Hesse 1980 (especially "The Strong Thesis of Sociology of Science"); also, recently, Jardine and Frasca-Spada 1997.

9. Note also how this position would not be any kind of "sociology of error": the Strong Programme's "principle of symmetry," which forbids asymmetrical explanations for those knowledge claims counted as true and those counted as false (as promulgated in Bloor 1991), is still fully in force, but simply bracketed.

10. The review has been singled out for favorable mention in the second edition of Paul R. Gross and Norman Levitt's *Higher Superstition* (1998, xii), perhaps precisely because it is, as they write, "astringent."

we are told, "led a simple family life and devoted all his time to research." Perutz goes on to endorse Pasteur's image as a "selfless seeker after the truth who was intent on applying his science for the benefit of mankind." According to Perutz, however, Geison claims "to have found him guilty of deception, of stealing other people's ideas, and of unsavory and unethical conduct." Indignation at this fuels Perutz's subsequent condemnation of Geison's book, a condemnation the basic strategy of which is to accuse Geison of having made scientific errors in his accounts of Pasteur's work. Such errors have led Geison to draw inappropriate conclusions regarding the moral character of Pasteur's science, and their clearing up correspondingly reinstates, in Perutz's mind, Pasteur's moral purity.

In fact, *The Private Science of Louis Pasteur* lends itself to this kind of attack because, as its title indicates, it operates according to an analytical demarcation between Pasteur's private and his public scientific faces. Geison studied Pasteur's private laboratory notebooks in order to compare the work reported in them with the accounts of that work that Pasteur presented in the public forum. In so doing, he particularly sought discrepancies between the accounts. It is these alleged discrepancies that Perutz attacks, since he assumes that they would, if genuine, imply a morally unacceptable picture of this "selfless seeker after truth." Geison, he asserts, does not understand the relevant science well enough to judge properly the relationship between the public and private accounts, and this in turn justifies Perutz's charges of "unethical and unsavory conduct" by Geison himself in an attempt to "drag Pasteur down."

Perutz's position becomes even clearer in his response (1997) to a subsequent letter from Geison (1997) that takes issue with the review. At a number of points, Perutz appears to ignore or willfully misunderstand Geison's explanations of where the review had failed to represent accurately the arguments of the book. That appearance deserves examination, however, because it indicates the deep-seated currents of cross-purpose that drive the rancor behind Perutz's reading of Geison.

Leaving aside the rather pro forma sneers at "relativism" that crop up in Perutz's review and response, one point is very clear. Geison notes in his letter that the "real disagreements [between him and Perutz] are epistemological, methodological, and ethical in nature." It can further be observed that the first two categories are subservient to the third. The epistemographical dimension of Geison's enterprise only becomes morally objectionable to Perutz when it is misunderstood as *epistemology*.

The logic of his reading appears to rest on the following assumptions. Pasteur, for Perutz, is the *fons et origo* of many of the benefits of modern scientific medicine ("the father of modern hygiene" and so forth). As such, he takes on the moral qualities attaching to those subsequent benefits. It is as if modern medicine, as a good, has been anthropomorphized, and then at least partly *identified* with the figure of Louis Pasteur. Consequently, when Geison represents Pasteur's scientific work as having been less than perfect according to the precepts of scientific methodology polemically advocated by Pasteur himself, or Pasteur as having been anything other than a "selfless seeker after truth" in some of his well-publicized exploits (the rabies treatment of Joseph Meister, or the celebrated anthrax trials), Perutz feels obliged to contradict Geison's picture in order to forestall a negative moral judgment on Pasteur and, by extension, on modern scientific medicine itself. Hence Perutz's frequent recourse to "what we now know" as a means of vindicating Pasteur's actions. Of Pasteur himself, Perutz says: "his greatness derives from a long series of fundamental discoveries which have stood the test of time and have much reduced human suffering" (1997).

The cash value of such a claim appears in Perutz's subsequent suggested rewriting of one of Geison's statements about Pasteur. Geison had made the epistemographical claim that, in studying the differential rotations of the plane of polarized light effected by its passage through solutions of left-handed and right-handed tartaric acid crystals, Pasteur had in fact recorded a slight difference between the two opposite-handed measurements (rather than them showing opposite deviations of the same extent, as he wished to find for theoretical reasons). Geison notes that Pasteur dismissed the difference as negligible, just as his published account of the experiment fails to discuss it.[11] Perutz pounces on Geison's description because he detects an implicit claim of Pasteur having cheated. Perutz's overall assertions on the point are somewhat inconsistent,[12] but his suggested alternative way to express the evidence from

11. See Geison 1995, 81–85, for the detailed account of this somewhat involved argument.

12. Perutz 1997. Perutz interprets Geison as saying that Pasteur's expectation, that well-chosen crystals would eliminate the residual difference, was "untrue" simply because Geison glosses Pasteur's words with the phrase "he now claims." This is an attribution to Geison of exactly the opposite view that Perutz had accused him of in discussing the same matter in the original review: "Later experiments [N.B.] by others proved Pasteur's explanation of the small discrepancy exactly right; but it seems that, because right and wrong would imply the existence of objective truth, they have been eliminated from the vocabulary of Geison's school of sociologists of science [*sic*]" (Perutz 1995). Thus Perutz alleges at one time that Geison rejects attributions of right and wrong, but at another

the notebooks that underpin Geison's account is clear enough: "Had Geison reported Pasteur's experiment truthfully, he would have stated instead: 'Pasteur found a small difference between the two measurements which he attributed correctly to the difficulty of separating the two kinds of crystals'" (1997).

The epistemographical point thus becomes very clear. Perutz ultimately rests his case on the *correctness* of Pasteur's assertions, regardless of the circumstances surrounding their utterance. The relevant circumstances are, of course, precisely those that prevented Pasteur from knowing what the results of "later experiments"[13] would show; that is, they are part of the temporal dimension of Pasteur's research endeavor as it proceeded. Geison, of course, was engaged in an epistemographical study in which such factors were essential.

Finally, it is worth noting another passage in Perutz's reply: "I accept the validity of Pasteur's work not merely because of his practical successes, but because his experimental results have never been falsified and their validity has been underpinned and given a strong theoretical basis by modern molecular biology. It is the hallmark of scientific genius to find the right answers even before the rationale for them becomes apparent."[14] Perutz's failure to understand Geison's work could not be made clearer. Nothing that Geison is criticized for saying is in the least bit inconsistent with a belief that Pasteur's answers, to the extent that they agree with present-day scientific belief, were "the right answers." The rightness or wrongness is simply irrelevant. Why is this point invisible to Perutz? Because "rightness" in the sense of truth-to-nature has, for him, collapsed into the notion of "rightness" in the sense of moral probity, in such a way that he takes the latter to imply the former.

EPISTEMOGRAPHY AND "PURE DESCRIPTION"

An obvious criticism of the idea that there is a clearly recognizable epistemographical core to science studies projects comes from the frequently voiced belief that normative elements are always implicated in science

time, on the very same matter, says that Geison makes an attribution of untruth to the relevant claim of Pasteur's. Calm deliberation was not at work here.

13. See quote in note 12, above.

14. Perutz 1997. See also the excellent letter, responding to Perutz's original review, by William C. Summers (1997), which stresses especially Perutz's failure to understand the nature of Geison's historical enterprise, a failure evidenced by Perutz's frequent citation of modern scientific beliefs and their vindication of Pasteur. Perutz's reply to the letter fails to take up the issue.

studies work—itself a point echoing the generally accepted contention in the philosophy of science that all observation is theory-laden.[15] As a point of principle, this is no doubt well founded; however, the matter is much less clear from a pragmatic perspective.

A well-known exchange several years ago in the journal *Science, Technology, and Human Values* examined the question of whether the science studies researcher can ever be above the fray when investigating scientific controversies. Three examples of attempts to study "disinterestedly" contemporary disputes in science displayed the phenomenon of the researcher becoming a captive of one or the other side, despite vigorous attempts to remain neutral (Scott, Richards, and Martin 1990). Taking issue with the skeptical conclusion derived by Scott, Richards, and Martin (the researchers who had conducted each of the studies), Harry Collins attempted to uphold the ideal of "disinterest" in studying scientific controversies as a methodologically valuable (as well as possible) approach, regardless of attempts by participants in the controversies to coopt the work of the "disinterested" researcher for their own use (1991). Collins's position requires that the in-principle neutrality of the science studies research methodology can be assured at the outset, and it is at that level of the argument that the theory-ladenness of observation might be invoked as a counter: use of some particular approach might of itself tend to favor one side over the other in the studied dispute (typically, in the cases considered by Scott, Richards, and Martin, the "underdog" [Martin, Richards, and Scott 1991]). For such reasons, the notion of an epistemographical (rather than an evaluative) account of a scientific controversy might appear chimerical.

However, this conclusion goes too far, ultimately because of the basic issues invoked by Collins. The pragmatics of research in many cases simply trump the epistemological critiques of the methods used. In the present context we have to do with the methodology of epistemography, that is, with an analysis of how one should go about producing truthful descriptive accounts of scientific knowledge and its making. There is no reason to despair, however, in the face of apparently looming regress or circularity, because (as Collins in effect indicates in his response to Scott, Richards, and Martin) the *process* is a matter wholly different from the *results*. The research process, or research methodology, is really a moral issue concerning how one ought to ask questions and seek answers; it is not a matter of propositions that claim a truth-value. One of the clearest

15. A classic statement of this idea is Hanson's (1965, 4–8) on the different "observations" of the sunrise by Tycho and by Kepler.

examples of such a view of methodology as morality may be found in Karl Popper's work.

Popper's demarcation criterion for science, according to which properly scientific statements are just those that are falsifiable, has received over the years a multitude of technical criticisms and purported corrections or improvements. With Popper's own professional connivance, most philosophical commentators have treated Popper's position as one aimed at providing technical rules of argument that will maximize the chances of creating, over time, reliable (verisimilar) scientific knowledge.[16] Perhaps, however, a more important response is to notice that Popper's epistemological and methodological prescriptions amount to moral rules concerning how people *ought* to go about making knowledge-claims in an honest and responsible fashion. In one of the pithiest expressions of his philosophical position, an essay called "Science: Conjectures and Refutations" (1972), Popper gives a biographical account of the origins of his crucial notion that the scientific status of knowledge-claims derives from their empirical testability—the possibility of their being shown to be false.

Popper's story had its roots in the latter part of the year 1919, a time when, he says, he was deeply impressed by the recent confirmation by Sir Arthur Eddington's solar-eclipse expedition of the light-bending prediction of Einstein's general theory of relativity. He had asked himself why this case seemed to contrast so starkly with the status of three other theories that were popular at the time, Karl Marx's theory of history and the psychological theories of Freud and Adler. In the latter three cases, confirmations of the theory could be found everywhere: "A Marxist could not open a newspaper without finding on every page confirming evidence for his interpretation of history. . . . The Freudian analysts emphasized that their theories were constantly verified by their 'clinical observations.' . . . It began to dawn on me that this apparent strength was in fact their weakness" (35). Einstein's theory was quite different because it could be confronted with observations or experimental results that differed from its predictions—and if in fact such results were not found, there was the mark of a good theory. Bad, nonscientific theories *could not* be falsified. Myths were of such a kind, perhaps based in observation, but not being liable to falsification. This logical argument, how-

16. Perhaps the most sophisticated critique and development of Popper's work was Lakatos 1970. Interestingly, Lakatos's elaboration of Popper can easily be seen as being as much imbued with moral precepts as Popper's work itself; nonetheless, it focuses tightly on technical philosophical issues for its structure and ostensible justification. Popper's *epistemology* as opposed to methodology is of course essentially experiential.

ever, also clearly amounted to a moral judgment. "There were a great many other theories of this pre-scientific or pseudo-scientific character, some of them, unfortunately, as influential as the Marxist interpretation of history; for example, the racialist interpretation of history—another of those impressive and all-explanatory theories which act upon weak minds like revelations."[17]

In his autobiography, Popper noted his awakening at the age of seventeen to the dangers of Marxism: "I realized the dogmatic character of the creed, and its incredible intellectual arrogance" (1976, 34). Such moral assessments were important underpinnings of Popper's philosophy of science and for its stress on a method of inquiry that would keep the inquirer honest.

For Popper, such considerations removed the teeth from many of the most damning philosophical criticisms that could be aimed at his falsification criterion for science. Indeed, the principal critiques were ones that he himself anticipated and addressed in his major philosophical statement, *The Logic of Scientific Discovery* (which first appeared in German in 1934). He knew that any empirical determination that a prediction made by a theory was false could easily be challenged, that the theory could always, in principle, be protected. The strategy for doing so was known as "conventionalism," and its potentially devastating consequences for a falsificationist view of science stemmed from an argument made at the beginning of the twentieth century by the French physicist-philosopher Pierre Duhem. Duhem had pointed out that there can never be a truly "crucial" experiment that will rule out some theory with logical necessity. The ingenious scientist could always place the blame for a theory's failure of an empirical test by locating the fault in some subsidiary part of the inferential system—perhaps it was not the theory under test that failed, but an unrelated auxiliary theory that the experimental system erroneously assumed to be true. For example, perhaps the understanding of the experimental apparatus's functioning was at fault rather than the theory that the apparatus was used to test.[18] By such means, an apparent falsification of a theory could always be deflected, perhaps generating in the process a new hypothesis about an unrelated matter

17. Popper 1972, 38–39. For another autobiographical account of the same issues, see Popper 1976, 41–44. For Popper's opposition to Marxist history, see especially Popper 1957. On Popper in the 1920s, '30s, and '40s and his attitudes towards Marxism, see Hacohen 1998.

18. Duhem 1962, 183–90. This basic point is nowadays known as the "Duhem-Quine thesis," honoring the philosopher W. V. O. Quine, who resurrected a form of it in the 1950s.

that would itself then be a candidate for falsification . . . and so on. This objection to Popper's idea did not discourage him, however. This is how he dealt with it in *Logic of Scientific Discovery:* "The only way to avoid conventionalism is by taking a *decision:* the decision not to apply its methods. We decide that if our system is threatened we will never save it by any kind of *conventionalist stratagem*" (1959, 82). Thus *moral* practice, rather than simply logical procedure, is crucial to Popper's conception of the proper method of science.

Perhaps it would be possible to represent any formal philosophical "method" for testing and evaluating theories in science in a similar way—to argue that all have at their core a moral practice that is accepted not because it is logically *necessary* but because it is in some way the most *honest* way of proceeding (recall Popper's scorn for those who held unfalsifiable theories such as Marxism; unfalsifiability for him went along with a morally insupportable refusal to adopt his "decision" to avoid conventionalism).[19]

Be that as it may, Popper's view of method in science is in this regard instructive: epistemography, like any purportedly descriptive enterprise, can always be convicted of failing to be value-neutral in some way, of smuggling into its procedures evaluations and interpretations that will color the empirical study with the prejudices of the inquirer. Like the possibility of using "conventionalist stratagems" in falsificationist science, such a charge could always be leveled at the epistemographical work of science studies. Popper wanted conventionalist stratagems to be outlawed, but of course this is infinitely easier to say than to enforce, since it is not even clear that completely avoiding them would be possible even in principle.[20] Popper should therefore be seen as in effect recommending that one always should *try* to avoid conventionalist stratagems, not that it is necessarily possible to *succeed.* The essentially moral

19. Francis Bacon's famous arguments in book I of the *New Organon* (1620) against existing philosophies and in favor of his new inductive method of inquiry are almost entirely and overtly moral in character, contrasting the "arrogance" and uselessness of some philosophies (e.g., the Aristotelian) with the diligence and humility required to do things his way. In this regard, it might be noted that his famous "Idols of the Understanding," which tend to mislead and to thwart true discovery, are described by Bacon in order that the reader, being aware of their dangers, will try to avoid them—but Bacon can suggest no technical "quick fixes" for doing so; he just tells people to watch out, to *try* to defeat them. It is a fundamentally moral injunction. Cf. Barnes and Bloor 1981, 46–47, for ironic remarks on "disinterested research." In the present context, it is interesting to note that Urbach 1987 provides a detailed and explicitly Popperian interpretation of Bacon's philosophy.

20. It was in a desperate attempt to improve on Popper's handling of the difficulties associated with conventionalism that Lakatos came up with his methodology of scientific research programs (1978).

nature of his philosophical analysis is all the more unmistakable because of what would otherwise be its clear technical shortcomings—shortcomings that Popper was much too good a philosopher not to see. So, similarly, the "neutrality" of the social researcher studying scientific controversies, or the "pure description" ostensibly sought by the epistemographer, must themselves be seen as *moral strivings,* not as achieved ends. The strict unrealizability of the ends does not devalue the worth of the striving.[21]

The descriptive ideal of epistemographical studies does not mean that science studies itself has no room for normativity, but it does mean that its various diverse endeavors are not defined by normative goals even in those cases where the science studies scholar wishes to include them. The epistemographical core of science studies may be utilized for a variety of ends, even including political ends, but such goals are not a defining part of science studies itself. Indeed, their possible deployment is itself a lively area of contention.[22] Where explicit recognition of the apparently rather banal need for descriptive accounts becomes important is in attempting a differentiation between description of what people do or have done and how we ought to *evaluate* their deeds. Geison's book holds that Pasteur did particular things that Perutz (and, in some moods, Geison himself) regards as improper or hypocritical. Perutz does not wish to see Pasteur as having done these supposedly unworthy things and uses as a major part of his anti-Geison armory assertions about what "science" has, in vindication of Pasteur, subsequently found to be true. It is easy to see that modern scientific beliefs are in fact irrelevant to an epistemographical account of Pasteur's work, and much the same point could be directed against many of the more indignant criticisms of other recent work in science studies (such as critiques of Collins and Pinch's book *The Golem*).[23]

Epistemographical accounts are, emphatically, not immune from criticism. They are of the same kind as any historical account, of nonscientific as much as of scientific subjects, and as fallible as any. However,

21. This is therefore a possible reading of Collins's position in "Captives of Controversy" (Scott, Richards, and Martin 1990) and corresponds to the chief argument of Collins 1996 (see especially page 232).

22. For discussions of such issues, see, apart from the "Captives of Controversy" debate, the special issue of *Social Studies of Science* (26 [1996], 219–468) on "The Politics of SSK: Neutrality, Commitment, and Beyond" as well as Radder 1998 and the replies that follow (332–48) by Vicky Singleton, Brian Wynne, and Radder. None of these debates could occur without acceptance of the possibility of some kind of epistemography.

23. See the new afterword to the second edition (1998) for consideration of some responses to the original edition.

they should be recognized as being immune from one sort of criticism, and that is the criticism that regards them as in some way *attacking* science. Some scholars do, no doubt, try on occasion to use epistemographical accounts as the foundation for attacking or undermining a current scientific practice or knowledge-claim. Those who do can certainly be criticized if they use evidence poorly in creating their descriptive accounts, but the critic must beware of using what might be regarded as undesirable inferences about science as grounds for the critique. When an historian or sociologist is attacked for telling stories that supposedly cast science itself in a bad light, the critic has in effect already accepted the proposition that epistemography can indeed provide accounts with the power to validate or invalidate scientific knowledge-claims. The die-hard defender of science would provide endless future hostages to fortune by admitting such an acceptance. Another scholar, by contrast, might find the epistemographical account perfectly well made and nonetheless reject the inferential steps that led the original author to draw *epistemological* conclusions about science. The wise opponent will usually do well to attack the inferential steps that link the epistemography to the evaluations, rather than attacking the epistemography itself, because if those steps are shown to be invalid, the status of the epistemography will be irrelevant to the case. To the defenders of science in the science wars, the epistemographical content of science studies should be utterly irrelevant. To practitioners of science studies, by contrast, epistemographical content represents the indispensable disciplinary core.[24]

24. This general (moral?) point raises the interesting question of whether Thomas Kuhn's *The Structure of Scientific Revolutions* (1996) should be seen as describing how the social institution called "science" works (epistemographically) or attempts instead to explain *why* science "works" (epistemologically). Certainly, much more progress has been made by those scholars who take the first road than those who have attempted the second in exploiting the possibilities of Kuhn's work.

Chapter 11

FROM SOCIAL CONSTRUCTION TO QUESTIONS FOR RESEARCH: THE PROMISE OF THE SOCIOLOGY OF SCIENCE

Kenneth G. Wilson and Constance K. Barsky

INTRODUCTION

The "science wars" controversy is sometimes presented as an "either-or" conflict: either scientific knowledge is culturally specific (as some social scientists claim) or it is truth and free of cultural contamination (as some scientists claim). But we believe in a middle ground. The scientific knowledge base surely includes both truth and culturally specific elements. The issue of culturally specific elements is apparent in early phases of a scientific investigation, when the existence of a phenomenon is still subject to challenge. But we believe that even scientific knowledge of high accuracy (such as the knowledge about the motions of the planets) still has culturally specific elements, but only at the present limits of that accuracy. For example, it is likely that French astronomers hold somewhat different beliefs about the errors of planetary data than English astronomers do, especially sources of error that are too small to have been resolved as yet. However, there is a problem that we must mention. What we, as scientists, refer to as "truth" involves some unstated assumptions, which trouble some sociologists.[1] We will compound this problem by raising new concerns for

1. E. B. Wilson Jr. has discussed unstated assumptions about scientific truth in his book *An Introduction to Scientific Research* (1990; see especially chapter 7 on classification), which was first published in 1952, but an example here is useful. The two of us regard the Earth as a natural object. But when questions arise about details of the Earth's atmosphere, we agree that the boundary between the Earth's atmosphere and the solar medium cannot be located precisely (to, say, millimeter accuracy) without a social agreement about how to define this boundary.

scientists and for sociologists of science about the nature of scientific truth.

While we will discuss the issue of cultural specificity and social construction in more depth, what really matters, we believe, is to identify root causes for the eruption of controversy about social construction. We propose that the controversy traces back to an internal problem of the sociology of science, that the sociologists were misdirected by their reading and interpretation of Thomas Kuhn's *The Structure of Scientific Revolutions* (1996). It is only now becoming possible, in hindsight, to recognize shortcomings in Kuhn's book that caused the misdirection. We will suggest a new interpretation of Kuhn's opus that could open up a new direction for future research in the sociology of science.

We begin, however, with a question regarding the version of "truth" that scientists use, a question that has been raised by sociologists of science and that underlies a legitimate but limited role for social construction in science. This question sets the stage for a discussion of Kuhn's opus and how it diverted the sociologists of science from examining some basic sociological issues. The next section outlines this question and the essential themes of this chapter in a bit more depth.

SCIENTIFIC JUDGMENT AND REPETITION OF SCIENTIFIC WORK

There is more to science than mathematical laws governing precise logical entities, such as the positions and velocities of planets. In addition to mathematical or experimental manipulations, scientists are constantly making judgments about their own work and about the work of other scientists. They use judgment to estimate errors of experiments or to assess the likelihood of mistakes in complex theoretical computations. When judging the claims of other scientists, they rely heavily on the reputations of those other scientists. But scientific judgments and reputations are not precise logical entities, arrived at through rigorous mathematical laws. Sociologists therefore legitimately ask: Could scientific judgments and reputations be the conduit for scientific knowledge itself to be compromised—at least to some extent—by social or cultural factors?

The scientists' stock answer is that, if so, there is no permanent compromise because any mistakes in judgment by any particular scientist will be found out when other scientists repeat his or her experiments and theoretical computations. However, experiments and theoretical computations are rarely repeated exactly. Instead, scientists are under relentless pressure to accomplish something new. They do this by improving on the experiment or computation they are repeating. They seek higher accuracy, or greater detail, than any prior experiment. But once these scientists claim to have unprecedented achievements, their experiments or calculations in turn need to be repeated. This sets up a process of unending repetition of experiments and theoretical computations: repetitions which in some cases have continued for *centuries*.[2] For example, the continuing prediction and observation of planetary motions has resulted in an accuracy that is now around one part in one hundred million for the positions of the inner planets.

In the sections that follow, we will argue that the accomplishments of centuries of repetition of scientific work, as in the case of the planets, considerably alters past discussions of what is scientific truth and what is culturally determined. We will pose a number of questions that have yet to be researched. Most importantly, we hope to identify a new locus for debate between sociologists and scientists. It will highlight a very fundamental discrepancy between the realities of recent history of society as a whole as many sociologists in general see it, which is a history of growing organizational complexity (especially after 1850), and the history of science as many scientists describe it, which continues to glorify the achievements of individual scientists (from Newton to Einstein) and to downplay the growth of institutions and their role in producing the far more powerful science of today than existed at the end of Newton's time.

The history of institutional development and change in science has a very different time scale than that for individual discoveries. Institutional changes in science are most profound on time scales of decades to centuries. This seems like an eternity to most scientists. In contrast, the results of a single experiment or computation can sometimes be obtained far more rapidly. We believe that sociologists who are legitimately concerned that individual scientific results (or technological devices) are deployed prematurely, before their limitations or hazards are understood, may wish to research the extent to which scientific or technological progress requires decades and centuries to achieve. Similarly, we believe that scientists need to ask whether they should give more credit and attention than now to institutional developments in science and technology that have required decades to centuries to achieve.

2. See the discussion of "progressive research programmes" in Lakatos 1978.

We suggest in addition that the history of the sociology of science itself merits a reexamination, with particular reference to the role that Kuhn's *The Structure of Scientific Revolutions* has played in this history. We suggest that this book needs a reinterpretation to take into account many new case studies from both the sociology of science and the history of science produced since it was published. The book is now inadequate without a reinterpretation because of its conflicts with details found in the case studies.

FROM "SCIENCE IN DEVELOPMENT" TO "SCIENCE AS KNOWLEDGE"

What particularly troubles the sociologists about the judgments of scientists is the seemingly abrupt transition they make from "science in development" to "science as established knowledge." When an area of scientific knowledge is in development, it is still controversial. Collins and Pinch (1993), and others cited in their references, have described the confusion that occurs as some scientists get deeply involved in executing variants on the experiments that have been done, while other scientists make judgments about which scientist to believe without getting involved themselves [15]. But then, a few years later, all the controversy is seemingly forgotten. Remaining is a list of experiments which scientists credit with helping to discover the "truth." The reputations of the scientists who performed these experiments are enhanced, while questions may be raised about other scientists whose experiments "failed."

We suggest that, from a sociological perspective, the transition from "science in development" to "science as knowledge" is a process that in many cases never ends because of the continuing need to repeat experiments. However, neither sociologists nor scientists have fully explored the nature of this process. In reality, we claim, the controversy over what experiments or which scientists to believe never fully disappears; the experiments open to question change (on time scales of decades to centuries—longer than most case studies, such as those of Collins and Pinch) as claims of progressively higher accuracy become the principal source of the continuing controversy.

For example, the uncertainties in the positions of the inner planets are now measured in parts in one hundred million. At these limits of accuracy, experimenters and theorists directly involved have considerable trouble deciding what experiments and computations to believe, just as they do when assessing claims for a major discovery.[3] If, let us say, scientists are arguing about the timing of occultations of stars by the moon on the time scale of milliseconds, then their timing to no greater accuracy than a second will likely be noncontroversial. But what is remarkable is that at the limits of accuracy, the science involved in planetary motions can sometimes be just as exciting (as well as controversial) as the science of a major new discovery. The classic example is the precession of the perihelion of Mercury, a tiny effect just barely measurable in 1850, which proved, as accuracy improved, to be a key test of Einstein's general theory of relativity. And even today, a frustrating aspect of planetary measurements is that they still lack (most likely by a large factor) the accuracy needed to detect gravitational effects on planets due to clouds of what astronomers call "dark matter." The galaxy as a whole is supposed to be profoundly affected by dark matter, but rough estimates of dark matter density make the density too small to be detectable yet in planetary motions.

We have said that mistaken scientific judgments will be found out when experiments or theoretical computations are repeated. But judgments are not mathematically precise entities, which means they can sometimes be wrong, even when made by the most skilled of scientists. This raises a question. How often do mistakes in scientific judgment actually occur? *We emphasize we are not talking about fraud here, but rather honest judgment calls that fall short, of the kind any scientist could make.* As far as we know, no statistics have been collected on this question.[4] We suggest that the long histories of improvements in accuracy in predicting planetary motions should offer a wealth of statistics on what fraction of claimed successful observations or computations had to be questioned later.

One interesting example to study might be the historical record of observations of Mars since 1750, along with the historical record of formulae (referred to by astronomers as "ephemerides") for predicting these positions. A large number of these observations have already been collected and published (Laubscher 1981, part 4). Predictions of planetary motions have been published in almanacs throughout the same period, and ephemerides from which such almanacs were constructed should

3. See Standish 1993. This paper attacks prior and controversial claims about the detection of a tenth planet beyond Pluto. The reference concerns the outer planets, for which measurements are less accurate than for the inner planets, but are still of far greater accuracy than in previous centuries.

4. See, however, Petley 1985. Petley's work is discussed in our second contribution to this volume, "Beyond Social Construction" [34].

be available too (Laubscher 1981, part 2, 189; Wilson 1980). Have the differences between prediction and observation steadily decreased since 1750, or were there setbacks from time to time, where accepted claims of accuracy had to be recanted? If there were setbacks, how many occurred, and what were their causes? How often did an individual scientist's judgments fail? How often did reputations turn out to be overblown?

CONTRASTING EXPECTATIONS OF SOCIOLOGISTS AND SCIENTISTS

Sociologists are aware of the social and cultural chaos of real life in a laboratory. If we had their background, we would find it hard to believe that this complexity does not often lead to unwarranted claims being accepted, once made. Surely, they could assert, judgments scientists use to support or reject a particular finding would prove fairly frequently to be inaccurate. Thus, we would expect overoptimism to be widespread regarding the accuracy of prediction of ephemerides.

Before we studied the sociological literature, we would have asserted that fellow scientists would be held in check by the knowledge that their claims of accuracy will be subject to scrutiny through repetition. We would have expected the number of cases of serious overoptimism, at least among scientists with a high reputation to protect, to be small.

Only an actual examination of the historical record can determine how often claims have proven to be overblown, and whether scientists with high reputations have their error estimates discredited less often by later repetitions of their research than scientists with mediocre or nonexistent reputations. *We do not know what to predict* as an outcome of such an examination. What is clear, however, is that the determination of errors in the predictions of planetary motions has been a complex task with many opportunities to overlook small effects (Wilson 1980).

There is a more profound reason for sociologists to study the histories of planetary predictions with care. We have searched quite extensively for sociological theories that could explain the repeated advances of the scientific enterprise in predicting planetary motions. As far as we could determine, *no such theory exists*. Given the present state of sociology, the achievement of the accuracies now routine in planetary predictions is apparently inexplicable, just as the complexity of the motions of the planets was inexplicable by physicists and astronomers before the development of Newton's laws of motion, or just as the generation of energy by the Sun for many billions of years was inexplicable before the discovery of nuclear reactions and the very large energies they can release.

But we must explain why the achieved accuracies for planetary motions are seemingly inexplicable within current sociology, and why we consider this inexplicability to be of great importance. In our opinion, it is not a trivial oversight, and certainly not a deliberate obfuscation.

ORGANIZATIONAL CHAOS AND SCIENTIFIC PROGRESS

One source of the sociological difficulty with explaining accurate predictions is that these achievements are the outcome of an overall enterprise spanning many independent organizations. There are academic departments of astronomy in many separate universities, here and abroad. There are observatories and telescope manufacturers. There are international organizations which define standards, such as standards for measuring time. There are scientific journals and funding agencies. To top it off, some of the most accurate ephemerides are now developed at the Jet Propulsion Laboratory (near Pasadena), which is in practice a part of the National Aeronautics and Space Administration—a complex federal bureaucracy.

The growth of all these independent organizations over time has been part of the overall growth of organizational complexity and bureaucratization that distinguishes modern society from medieval days. The problems this transformation of society has caused are a central concern of sociologists. Sociologists, we believe, have very legitimate reasons for asking how the difficulties of this transformation have affected the production and verification of scientific knowledge.

Sociologists know a great deal about the operations of complex human organizations, such as those within which scientists now operate. They know that complex organizations and bureaucracies do not always function as advertised. These organizations are rent by power struggles and hidden agendas and financial crises. They may hire or promote inappropriate candidates for key positions. They often avoid cooperation with other organizations, based in part on the "not invented here" syndrome. Moreover, organizational actions often have unintended consequences.[5] To make matters worse, the scientific enterprise has no one to

5. One example of unintended consequences that sociologists cite is a New York City law which required owners of deteriorating residential buildings in low-income areas to bring their properties up to minimum standards. The result was not improved living quarters for the poor, as was desired; instead, deteriorating residences were either abandoned or converted to nonresidential use (Giddens 1996, 5–6).

direct it; in business parlance, it has no chief executive officer who could impose some kind of order and direction on the operations of the enterprise as a whole.

What we would have expected, from a sociological perspective, is that many of the independent organizations making up the astronomical enterprise would long ago have started working at cross purposes, making further scientific progress impossible. The true errors in prediction of planetary motions today, according to our sociological analysis, should not be much better than they were in 1750, when the astronomical enterprise was already complicated enough to have retarded further progress on predictive accuracy because of organizational chaos and unintended consequences.

We believe sociologists should ask specific questions such as why failures like the initial problems with the Hubble space telescope are not so commonplace that they could not be overcome without a total replacement. They should ask why the training of young scientists does not constantly fall behind as conservative faculty block curriculum changes necessary to incorporate scientific advances. They should ask why funding for science is not more often misdirected for political ("pork barrel") reasons. In response, scientists would likely claim that these failures are mostly limited to less successful components of the scientific enterprise, such as university departments with poor rankings; but is this true?

We believe that there need to be extended discussions between scientists and sociologists of science as to whether the production of very accurate predictions by the astronomical enterprise constitutes a paradox that is inexplicable according to the present state of sociological knowledge, or whether the sociologists have a valid way of accounting for these successes.

A KEY TOPIC: CONTINUING IMPROVEMENT IN PLANETARY STUDIES

There is another aspect of planetary astronomy that we believe should be useful for both scientists and sociologists to explore. As predictions and observations of the planets were repeated, there were repeated improvements made to observational techniques, to telescopes and other instruments, and to methods of theoretical computations in order to achieve increased accuracy. For example, there were many stages of continuing improvement between Galileo's hand-held telescope (little more than a spyglass) and the Hubble or Keck telescopes of today. There were even theoretical revolutions, such as Einstein's theory of relativity, which must be used to achieve current accuracies (especially Einstein's prediction of the bending of light by the Sun's gravity).[6]

There also were changes in the organizations that generated or made use of these improvements: they added more faculty or employees, and new roles emerged for these faculty or employees (such as new subdisciplines of astronomy or physics, or new technical or managerial roles). Educational requirements increased repeatedly for some of the most critical roles. The companies and departments that help build telescopes today have far greater capabilities than the organizations that helped with earlier telescopes.

The crucial role for scientific progress of continuing improvements, especially of the accompanying organizational changes, is seldom recognized. But this raises new questions for research. For example: How were repeated increases in funding achieved when they were needed to support the continuing improvement? Was it helpful that continuing improvements steadily increased the usefulness of the research beyond the specific focus of planetary motions, for example, that telescopes could see farther into the sky and that Newton's laws could be applied to more complex situations on Earth as well as in space?

CONTINUING IMPROVEMENT IN THE SOCIOLOGY OF SCIENCE?

We believe that sociologists of science face just as much sociological pressure to repeat and improve on prior work as do other kinds of scientists. In response, they have built a growing literature of case studies and theoretical analysis, similar to what many branches of the natural sciences are doing.[7] At its best, this literature is of high quality, as one of us can attest from personal experience: A case study by Andrew Pickering (1984) traced the development of the quark hypothesis in elementary particle physics. This research field engaged one of us (kgw) during the time period of Pickering's study.

The sociology of science is still a young field: its founding traces back only to the early 1900s, in contrast to the astronomy of the planets, which was over two thousand years old at the time of Newton. Can one

6. E. Myles Standish, private communication to the author (kgw).

7. There has been a buildup of case studies in the history of science also (Norton and Gerardi 1995).

expect the same level of accuracy in the sociology of science as has been achieved in the astronomy of the planets? Pickering, in the spirit of continuing improvement, sought to clarify the controversy that followed Kuhn's *The Structure of Scientific Revolutions*. He makes this clear in his last chapter, when he refers to Kuhn's concept of paradigm shifts and finds evidence for Kuhn's characterization of conflicting paradigms as "incommensurable,"[8] but he also finds that the controversy Kuhn expected following a paradigm shift can be absent, as during the transition within high-energy physics, around 1972, to the current theory of quarks, "quantum chromodynamics."

We will return to Pickering's work later. Meanwhile, we will highlight some other serious but subtle deficiencies that can now be recognized in Kuhn's own work.

REINTERPRETING KUHN

There are several areas where Kuhn's book is incomplete, we believe. The first is that it contains no overview of the sociological ideas offered therein. The ideas in it have to be dug out, and this is not easy to do. The second is that although Kuhn offers a number of novel hypotheses about the sociology of science, he offers inadequate proof that these hypotheses are correct. He provides anecdotal evidence for many, which is sufficient in some cases to justify their further examination, but the evidence as given is incomplete and in some cases Kuhn's hypotheses seem too implausible to survive detailed investigation. All Kuhn's hypotheses should, we believe, be regarded as tentative and incomplete until more thoroughly tested than they have been to date. Another major problem, we suggest, is that these hypotheses are too rigid; Kuhn allows too little flexibility for adapting them to the contexts historians or sociologists have uncovered in specific case studies.

What is the essence of Kuhn's book, from a sociological perspective? We suggest that it concerns the examination of the culture of a scientific discipline or subdiscipline, with that culture comprising material artifacts (such as books, articles, and laboratory notebooks), sociological roles (such as pioneer—a proposer or supporter of a new paradigm, or skeptic—a defender of an existing paradigm), socialization processes for new recruits (formal courses and Ph.D. training), and belief systems (the

8. For clarification of Kuhn's work, including his use of the term "incommensurable," see Hoyningen-Huene 1993.

paradigms themselves). The key value of Kuhn's approach is that these cultures can take radically different forms, and he offers the beginnings of a taxonomy (some would call it a typology or a classification). He proposes a taxonomy based on four distinct stages of science, which he labels "the route to normal science," "normal science," "extraordinary science," and "scientific revolution," with each stage exhibiting a distinct form of culture.

To support his approach, Kuhn proposes a distinct set of sociological characteristics for identifying each of his four stages, though it would have been helpful if he had summarized his characteristics in the form of four checklists. Instead, one has to go through Kuhn's book paragraph by paragraph to locate nominations for such checklists. For instance, there are a number of characteristics proposed for the route to normal science, all mentioned in just four paragraphs (two on pages 12 and 13, two more on pages 15 and 16): large numbers of facts are collected from the "wealth of data at hand"; the results are hard to interpret and sometimes incorrect; multiple schools of thought develop, each choosing a different subset of the claimed facts to explain; researchers write books which build from scratch an analysis of a subset of the data, being unable to take any facts or theory for granted.

A striking feature of the checklists, once compiled, is the profound differences between them. For example, in contrast to the case of the route to normal science, during periods of normal science there is a shared paradigm that is learned from textbooks. Research articles take the material in textbooks for granted. Instead of random collecting of facts, research is now confined to a "relatively small range of relatively esoteric problems," and "the paradigm forces scientists to investigate some part of nature in a detail and a depth that would otherwise be unimaginable." (These characteristics are among many that he proposes for normal science, from pages 23 to 43.)

Some of the characteristics Kuhn proposes arouse skepticism, at least from us. One example is his characterization of normal science as "mop-up work" (24). This label seems inappropriately weak to describe the complexity of the research on the planets following the publication of Newton's *Principia*. It is this label, and his justification for it, which, we fear, has diverted the sociologists of science away from investigating the growth of both scientific and organizational complexity over long periods of normal science. Scientific revolution, as a result of Kuhn's novel characterization, has attracted more attention from the sociologists of science than the long-term growth of normal science has.

For example, Pickering never refers to Kuhn's discussion of normal science, or the route to normal science, yet one can find indications of both stages of science in his book. In particular, in his closing summary, he comments that the research he described from the 1960s focused on "the most common processes encountered in [a high-energy physics laboratory]" (410), that is, picking up facts readily at hand. After the emergence of a theory of quarks, the focus changed to "rare phenomena" (410)—corresponding to Kuhn's "small range of relatively esoteric problems."

Pickering also does not identify what is (to us) a plausible nominee for a paradigm shift that appears in his case study: the first proposals of the quark hypothesis. The proposals were made independently by Murray Gell-Mann and then graduate student George Zweig in 1964. Between 1964 and the early 1970s, Pickering reports an intense controversy over the quark hypothesis, just as one would expect from Kuhn's description of a paradigm shift. And what follows the first quark papers cannot be described as "mop-up work," because the central and most complex developments in Pickering's case study occur after the quark papers and lead up to the proposal of a formal theory of quarks (the aforementioned quantum chromodynamics) in 1972. However, despite these examples, we believe there was no reason for Pickering to give more attention to Kuhn than many of his colleagues in sociology or in history now do.

What is needed, we believe, is to ask how Kuhn's analysis fares when applied to a large set of technically and historically detailed case studies from the accumulation that now exists. Which of Kuhn's hypotheses help make sense of a considerable number of these case studies? Which ones can be quickly discredited? Which ones could work, but only with modifications?

What is particularly difficult, even for some admirers of Kuhn, is matching Kuhn's rigid checklists to the chaotic contexts that the case studies reveal. He leaves no room for expert judgment in applying the checklists to specific cases. Therefore, the rationale for each hypothesis should be clarified in order to determine, in each case, whether a characteristic is unavoidable or might only occur in some cases.

Take, for instance, the claim that books written on the route to normal science can take nothing for granted. This statement seems extreme; there could be some unstated assumptions that everyone in a field shares, despite the existence of competing schools of thought, even if there are not enough to justify a textbook. In Pickering's case, the situation was more complex; physicists could not have worked successfully in high-energy laboratories without taking for granted the more established areas of physics covered by textbooks, from electrodynamics to quantum mechanics. Nevertheless, on specifics of the substructure of protons and neutrons there were competing schools of thought in the 1960s, as well as some other of Kuhn's characteristics of "the route to normal science."

Because of the problems we have outlined, we suggest that the time has come to reanalyze Kuhn's book as an initial, highly tentative proposal that a taxonomy of scientific cultures could be constructed. We suggest that sociological checklists (with considerable flexibility in their use) be constructed for each kind of culture represented in the taxonomy. We suggest that there would be more than four entries in the ultimate taxonomy. (One of Kuhn's omissions, ironically, is the taxonomy-building stage that many scientific disciplines have gone through.) We suggest that the growing accumulation of case studies of scientific episodes that give full attention to technical as well as historical detail (such as Pickering's example) be analyzed in order to determine how Kuhn's initial hypotheses need to be revised to move closer to a working taxonomy, perhaps using the history of the building of major taxonomies of the natural sciences as a partial guide as to how to proceed. We suggest that Kuhn's description of normal science needs major reworking to take into account the phenomena of decades of continuing improvement. We suggest, finally, that the proposed taxonomy is unlikely to spring into existence rapidly, but rather by successive approximations, much as the improvements in planetary predictions has proceeded largely by successive improvements.[9]

CONCLUSION

We believe that the current controversy over the social construction of scientific knowledge hides a more interesting prospect: that the sociology of scientific knowledge has the potential to become a major source of exciting opportunities for research. It could branch out from the current focus of controversy to address other issues as well, from the growth of organizational complexity in science to a renewed interest in the work of Thomas Kuhn. But since excitement in science often arises where one least expects it, especially in a research field as young as the sociology of science, we do not claim that we have identified more than a small

9. Frank Sulloway (1996) has made an initial effort to catalog a number of episodes of scientific revolution.

subset of the future directions in which scientists and sociologists of science in collaboration could find rewarding pursuits.

ACKNOWLEDGMENTS

We are grateful to all the participants in the recent science wars controversies: to the sociologists of science and their critiques about the truth claims of science as much as to scientists who have defended their truth. We owe specific debts to Kurt Gottfried,[10] George Smith,[11] Jed Buchwald, Sylvan Schweber, John Burnham, Trevor Pinch, Andrew Pickering, and Seymour Sarason for discussions and assistance. One of us (kgw) has benefited from prior discussions with Arnold Arons spanning several decades.

10. Kurt Gottfried coauthored a precursor for this article (Gottfried and Wilson 1997).

11. George Smith has been researching the inauguration of successive approximations by Isaac Newton, which is a specific version of the continuing improvement that we discuss (Smith, forthcoming).

Chapter 12

A MARTIAN SENDS A POSTCARD HOME

Harry Collins

Caxtons are mechanical
birds with many wings

And some are treasured
for their markings

They cause the eyes to melt
and the body to shriek without pain

I have never seen one fly but
sometimes they perch on the hand

—Craig Raine, *A Martian Sends a Postcard Home*

STRANGENESS AND HOSTILITY

The poem above describes a Martian's view of earthly things seen for the first time; it is a stranger's view. There has to be something strange about the sociologist's view of science too. It is the job of the sociologist to estrange him or herself so that those things that are taken for granted in the native society—those things that seem just common sense—become things that require explanation.

Accounts of the everyday life of a society from an estranged perspective can seem threatening. Consider how the sociologist might talk of religious beliefs. Imagine the sociologist trying to understand how it is that south of a certain line in a certain country it is generally believed that a certain kind of religious ceremony can turn wine into blood, whereas north of that line it is generally believed that no such thing occurs. The sociologist's account would be in terms of the histories of the respective beliefs; the processes of socialization through which the young are inducted into the beliefs; the separate social networks which reinforce the separate beliefs; the divergent conceptual structures in which the separate beliefs are embedded; the day-to-day actions in which the conceptual structures are expressed; the economic and political interests which drive the upholders of these beliefs apart; and, perhaps, the social psychology of cultural polarization.

What would *not* be taken into account by the sociologist is the matter of whether the wine really does turn into blood. Instead of accepting them at face value, the sociologist would "relativize" the beliefs—show how they relate to their respective societies. That is why the sociological account might seem threatening. To the believer from the southern part of the country, to ask, "Why do you think the wine turns into blood?" or to suggest that the southerner, given a different upbringing or experiencing different sets of social or political forces, would believe something different seems in itself to confront the belief. The believer believes the wine turns into blood because it turns into blood; that is what being a believer means [14].

But giving a sociological explanation of a set of beliefs and actions is not always seen as an attack. Suppose I were to provide a sociological explanation for the beliefs and actions of Nazis before and during the Second World War. Suppose I were to say, along with certain (Jewish) analysts,[1] that the social forces on individual Germans under the Nazi regime were so great that it was almost impossible for them not to come to see certain classes of people as subhuman. Furthermore, suppose I were to agree with these analysts that what would count as barbarity in respect of humans seemed to Germans, in respect of *sub*humans, merely the working out of mundane bureaucratic rationality. In that case the sociological explanation would appear to be a defense of the Nazis and an excuse for cruelty. It would seem to be a version of the Nuremberg defense ("I was only following orders")—in this case, "I was unselfconsciously acting as an ordinary member of my society."

It appears, then, that whether a sociological explanation is seen as an attack on or a defense of a set of beliefs and actions is itself relative to the valuation of those beliefs and actions within the group considering the explanation. Sociological explanations are seen as attacking those who hold positively valued beliefs and engage in the corresponding ac-

1. Hannah Arendt and Zygmund Bauman.

tions and defending those who hold negatively valued beliefs and engage in those corresponding actions.[2]

The relativity is seen still more clearly because the viewpoint changes as the viewer changes. The analysis of Nazi beliefs and actions would have seemed insulting to a doctrinaire believer at the height of the movement; to such a person, attacks on Jews were justified by theories of race and the Jewish financial conspiracy. No further explanation was needed. Similarly with science: to a scientist, sociological explanations of dominant theories seem to involve a misunderstanding, whereas similar explanations of "fringe science" or "pathological science" seem reasonable. Thus, the perception of the sociology will vary with the changing fortunes of the science that it tries to explain.

This relativity of the degree of hostility of sociological explanation suggests two things. First, it suggests that analysis is not necessarily hostile just because it is sociological. Practitioners of sociology of scientific knowledge (SSK) argue that what they do is not intrinsically challenging to science. Though it may be strange-looking and though it may be widely perceived as hostile, there is nothing in the logic of the sociological approach that actually makes it hostile. This is not to say that there are not some kinds of social analysis and some analysts that *are* intentionally hostile to science, nor that there are not some aspects of the behavior of scientists to which nearly all social analysts are in fact hostile. But, over a large range, the extent to which sociological analysis is perceived as challenging is orthogonal to the intentions of the analyst; sometimes intention and perception will match, and sometimes they will not. And sometimes the goodness of fit between intention and perception will change from place to place and time to time for reasons that have nothing to do with the analyst.

The second point is that sociological analysis and evaluation are also orthogonal. This is most easy to see in cases with a high moral charge. For example, though one might agree with those analysts who think that what would count as savage behavior in respect of humans seemed to some Germans merely the working out of mundane bureaucratic rationality, the actions remain just as vile and unforgivable; the negative evaluation remains just as strong. In the case of science the same principle holds with the opposite consequence: sociological

2. And this is why it is not the case that sociology of scientific knowledge is always perceived as supporting the underdog. The consensus view will be that the sociological position defends the underdog, but if underdogs believe they are right, then the sociological explanation of belief seems to them to be just as much an attack on their ideas.

explanation is not the same as disparagement; as far as the analyst is concerned, the value placed on the science remains what it was (in most circumstances).

Unfortunately, it is true that an analysis that is widely perceived as hostile, irrespective of the analyst's intention, can be damaging to its object. How should one react to this? One might stop such analyses or one might try to change the widespread perception. It is toward the latter that I believe the actions of both sociologist and interested scientists should be directed.

ARE SOCIOLOGISTS JUST LIKE MARTIANS?

Sociologists do not come from Mars. But from mere proximity to a social or cultural group, understanding does not follow. I do not understand Catholicism nor Anglicanism even though my partner is a lapsed Catholic and I went to a school run by the Church of England and sat through prayers every morning until the age of eleven. ("Give us this day our daily pret" [as I heard it]. "There is green hill far away without a city wall." What were they on about?) I still cannot make any sense out of the idea of "gentle Jesus meek and mild," nor "the father, the son, and the holy ghost." What is a "holy ghost?" To do some sociology of the type that most practitioners of SSK endorse I would have to try to understand these religions in a deeper sense and to get some feeling for what these sayings and icons inspire in believers. Sociologists need to know more than Craig Raine's Martian. The job of sociologists is to understand first and then estrange themselves later.[3] It is the job of sociologists of science to gain as much of every kind of understanding of science as they can. But having done that they must estrange themselves once more so that they can analyze the world from the peculiarly sociological perspective. This is where sociologists differ from the scientists they study. The scientists, *as scientists,* have no reason to estrange themselves from their practices, except that it might be edifying, as in the relationship of art to ordinary life.

Some critics of science studies assume that the analysts have never held any relationship to science *except* that of the Martian. This is the burden of Alan Sokal's hoax, and of remarks such as that of the apoplec-

3. Henceforward by sociologists I will mean "interpretivist" sociologists—interpretative sociology is favored by most sociologists and historians of scientific knowledge. Compare Latour and Woolgar 1986.

tic Duke University scientist who complained that the authors of *The Golem* had "never been within a quarter mile of an experiment" (Evans 1996). Science studies has grown so big that I cannot possibly evaluate the validity of this assumption for anyone except the small group who are happy to describe themselves as working in SSK. The members of this group *do* start with an everyday familiarity with science gained through a variety of routes. Most have been trained in aspects of social science which are scientific: they have been trained in the language of hypothesis and test and know what it means for data not to fit a theory. Some have done experimental or other empirical research in the social sciences of a type which is quite close to natural scientific work. Some have been trained in the natural sciences, in one case to the level of postdoctoral fellow in high-energy physics and in another to the level of published author working at a prestigious center of radio-astronomy, while the postdoc currently working with me is still publishing in the field of theoretical astrophysics in which he is simultaneously doing sociology research. Certainly, all researchers who have done empirical work will have developed at least enough competence to understand what drives their scientist respondents in the particular fields in question. All those working in SSK are familiar with science in a way that, say, I am not familiar with the meaning of Catholicism or Anglicanism. Furthermore, those who practice SSK warrant their work by reference to its empirical validity and replicability; they simply *are* members of the scientific community. Were this not the case SSK analysts would not be able to do their empirical work because they would not be able to talk sensibly to their respondents. Thus in writing about science the way they do, SSK analysts are not failing to understand the scientist's viewpoint—into which they can switch at a moment's notice—they are writing from a different viewpoint. Part of the social scientist's repertoire of skills is to be able to switch viewpoints in this way.

This difference in viewpoint has come out clearly in the interchange in *Physics Today* between Mermin and Pinch and myself (Mermin 1996a, 1996b; Collins and Pinch 1996; Mermin 1996c; Collins and Pinch 1997; Mermin 1997). Mermin complains that we are not interested in the physical truth of the matter. He is right: we are not interested in the physical truth of the matter when we are writing as sociologists. But we do know what scientific truth means to scientists; we are scientists and guard our own truths just as jealously as we expect physical scientists to guard theirs.

AN EXAMPLE: RESEARCH ON GRAVITATIONAL RADIATION

I will try to illustrate some of these points through my current research on the science of gravitational radiation detection.[4] Since the 1960s attempts have been made to detect gravitational waves by observing their putative effects on resonant masses. In the early 1970s it appeared for a time that the waves had been seen. The energies implied by the apparent strength of the waves combined with the apparent levels of sensitivity of the detectors were, however, about nine orders of magnitude too high to fit with current astrophysical theories. By 1975 many further experiments had been done, and certain clever statistical analyses had been carried out; by this time the positive results were no longer believed by more than one or two people in the world. Based on interviews I conducted in 1972 and 1975 I wrote several papers analyzing this episode from a sociological point of view and arguing that the series of experiments would not have been seen as decisive had it not been for the strong leadership of one scientist; he convinced nearly everyone that they should interpret their negative results as revealing the nonexistence of gravitational waves rather than their own incompetence.[5] Throughout this period, and throughout my analysis, I did my best to remain neutral as to the existence of these "high fluxes" of gravitational waves.

Two decades later I am revisiting the field of gravitational radiation research and doing my best to regain my neutrality—something that I set aside when my sociological interest in gravitational radiation ceased.

The field has changed somewhat since I left it. The only working state-of-the-art detectors are still resonant bars, but they are now cooled to liquid helium temperatures or below.[6] Nevertheless, according to standard theories of sensitivity, the feebleness of the gravitational radiation emitted by foreseeable astronomical events means that these bars are still two or three orders of magnitude too insensitive to detect anything except events that might take place a couple of times per century. From

4. For an easily accessible introduction to this project see my Web site: http://www.cf.ac.uk/socsi/gravwave.

5. It turns out that the physicists involved have now read these papers. Most of them think they are accurate and fair.

6. Joseph Weber, who made the original positive claims, believed (until his death in September 2000) that his room-temperature bars at the University of Maryland were more sensitive than the cooled bars because they were attached to better transducers.

time to time, however, some resonant bar groups have reported what seem to be gravitational radiation signals.

A typical state-of-the-art resonant bar antenna might cost a half-million dollars to construct. From the mid-1980s a new technology for detecting gravitational radiation has been developed. This is based on extremely large laser-interferometers. In 1992 Congress agreed to fund the so-called Laser-Interferometry Gravitational-Wave Observatory—LIGO. This consists of an interferometer in Washington State and another in Louisiana, each with arms 4 kilometers long. The cost of this project is likely to exceed $300 million. (Smaller versions are being built in Europe.)

I have just published a long draft paper on a small aspect of this field—how the propensity to publish speculative results emerging from the resonant bars differs between an American group and certain non-American groups and how this has been affected by the funding of LIGO (1998a).[7] I have visited the relevant experiments several times since I reentered the field and have recorded many hours of discussion with the scientists; I have also attended a number of relevant physics conferences and recorded the discussions of scientists with each other and with me. The foreign groups are willing to publish results that might be gravitational radiation but cannot be clearly said to be so; the Americans are unwilling to do this.

My paper consists of about forty closely typed manuscript pages, much of it devoted to explaining techniques of detection and analysis and how different interpretations of what might be signal and what might be noise can be supported by technical arguments.

A small section of my paper turns on a sociological analysis of the different national groups. I relate the non-Americans' willingness to publish results while they still contain a substantial measure of speculation to, among other things, a collectivist as opposed to individualist ethos about how findings should be interpreted. The non-Americans consider the determination of the significance of a claim to be the responsibility of the community of scientists rather than the individual laboratory; the Americans see it as a matter of honor that each individual laboratory puts out results that are eventually agreed to be correct and significant.

I then go on to describe the different pressures experienced by the groups. I try to show how the differences make sense in terms of, among other things, the funding climate in the respective nations. LIGO is con-

7. This paper is also available at the Web site mentioned in note 4.

tinually at risk from the bad name that might be given to gravitational radiation research by a spate of incorrect results. US resonant bar research is supported by the same National Science Foundation that supports LIGO, and the American group needs continual refreshment of its grant support from NSF. Thus, the American group, insofar as it experiences pressure from funding sources, experiences it as a pressure toward being conservative. The non-Americans, being funded as civil servants, experience no such pressure; on the contrary, they seem to feel the need to put out publications so as to justify their work.

In my paper, all this is documented with quotations. I also include substantial caveats pointing out that the conservatism I discuss should not be seen as directly caused by the funding pressures; my paper, I explain, is not trying to explain the choices people actually make, but to set out the pressures within which they work. I explain that my business is not motives but social and cultural patterns. I explain in the paper that this is the sociological perspective on the matter and that I would not necessarily expect it to be shared by the physicists.

I sent out the draft paper to the principal scientists involved, asking permission to use the quotations from their interviews and asking them to comment on whether they think it is accurate and fair. I will concentrate on one reply from a scientist who is central to the field.

In his response to my paper this scientist expressed himself happy with the quotations I had taken from our conversations. He also said:

> On the whole the article is entertaining and gives an accurate sense of the divisions of viewpoint within the community.

But

> I'm afraid that we're poles apart on the "adventurous" vs. "conservative" terminology. As you'd expect, I'd claim that more accurate terms might be "incorrect" vs. "correct," or "irresponsible" vs. "responsible," or "reckless" vs. "professional," "bad science" vs. "good science." This has nothing to do with geography, or my connection to the establishment. Rather, the fundamental canons of scientific research are at stake. A first class experimentalist cannot leave some bizarre result for others to analyze. They can't. It's his experiment, and only he can have access to the apparatus and data with enough understanding to do the job. It is his responsibility to keep at it until he kills the spurious result, or understands its origin and has very strong proof of its reality. (Say at the 7-sigma level.) Until then, nothing should be said.

I would say that this amounts to a jealous guarding of what I describe as the typically American norm of individualism: "it is the job of each individual laboratory to establish the truth as far as is possible before releasing anything to the community." I do not misunderstand this point—I understand it a lot better than I understand the Catholic mass—but I want to point out that it is a cultural value, something that is learned as one grows up within the community of scientists. If my analysis is correct, if one grew up among other groups of scientists one might learn a different value. But we can see that the very act of pointing out that there is a difference in values looks like an attack on the value; a value that is less than universal seems a lesser value by that fact. Universality lends a value the stature of an absolute as opposed to the quality of something that is colored by mundane and temporary circumstances. Discussing differences between the values of different groups seems, just as we saw in the case of different interpretations of the Catholic mass, to take away the sense of the absolute.

Once more, I want to insist that my own position, as a scientist, on the dimensions of judgment I have described, cannot be read off my analysis. I may have a preference, but it is not germane to this debate.[8] Professionally, I do everything I can to try to remain neutral on questions of the physics of gravitational radiation research, both in terms of findings and in terms of the norms of practice. As we have seen, however, neutrality of this sort looks like bias. My American respondent goes to the heart of the matter with the following warning: "Watch out with your objectivity, you'll catch a lot of flak. There is no way to do amoral analysis of this methodological split, just as some moral perspective is needed in discussing the methodology of Attila." Right at the center of the methodology of sociology and history of science is the prescription to do "amoral analysis," that is, to leave the scientists to decide on the truth while the analyst records the argument without taking sides. If there is no way of doing amoral analysis, then the sociology of scientific knowledge is dead.

WRITING HOME

It was all a lot simpler when the Martians sent their postcards back to base on Mars. In the first two decades of SSK no one took much notice

8. Just as I read physicists' preferences from the way they do their physicist's work, anyone can try to read my preferences from the way I do my sociologist's work.

of what was written except other historians and sociologists; under these circumstances the Martians could do their work in relative peace. But SSK had to grow up some time and join a wider community. The fact is that, being members of the scientific community, Earth is the natural home to practitioners of SSK. These debates are the symptom of SSK coming home after a long voyage. We have picked up the mannerisms and ways of speaking of outlanders during our long voyage; what is worse, we won't abandon them.

How then are we to learn to talk once more to the scientist earthlings? How, for example, can I continue to talk to my professional friends in the gravitational wave community when they are reading attacks on me in *Physics Today?* As my respondent put it in a phrase recalling Arnold Schwarzenegger's *Terminator* role: "We can argue more over beer, but cover your butt or else Mermin will be back" and expanded in a later commentary: "[M]y warning was that with your present emphasis, you're about to become a lightning rod again. . . . [Y]ou need to be careful about your credibility. . . . [I]t's the same issue that the experimentalists face. A scientist stakes his reputation on his results. If they are silly, half-baked, or incorrect, people stop listening, and it's almost impossible to get back one's good name ever again." The easy solution is to trim and duck for cover. I could take out the "sociological" section of my paper; a few cosmetic changes and it would become a straightforward description. I might actually do this, but it cannot represent a long-term solution. In the end the reasons are moral. To do amoral analysis of physics I have to do a highly moral sociology—a sociology the very existence of which is an affirmation of academic freedom, however uncomfortable this is for the analyst. This is an old story: as soon as one takes a defensive stance in the face of attack, one legitimates a witch hunt.[9]

In any case, though SSK wants to come home, and needs to come home, its voyage of discovery will have been a waste of time if homecoming means casting off all outlandish habits. The point of SSK is to add some variety to the way we look at science. Contrary to what some writers think, SSK is not in the business of vying with science for a claim on physical reality; at the heart of its method of amoral analysis is a determination to leave physical reality to the physicists. But this is not

9. The *Chamber's English Dictionary* definition of a witch hunt, by the way, is: "The searching out of political opponents on grounds of alleged disloyalty to the state etc.; also applied to any similar non-political search or persecution of a group or individual."

the same as being a pale reflection of physics or a historical handmaiden.[10]

In the long term what we have to do is carry on talking to each other in the way we are talking in this book so that suspicion dissolves and the point of the estranged perspective of the sociologist becomes less misunderstood. Probably the most useful point for discussion in the immediate future is the way the history is to be treated, whether retrospectively or contemporaneously. SSK, if it is to remain true to itself, will always treat the history of science differently from the way the scientist would treat it, and this needs talking through. But even in the longest term the perspective of the sociologist is meant neither to replace that of the physicist nor to reflect it; it is just a different perspective. One might say that while Earth is home, the view will always be colored by the ruddy glow of Mars.

10. Gottfried and Wilson's critique of SSK (1997), which complains that it does not continually update itself in terms of the current state of physical knowledge, misses this point. SSK is not a kind of Dow-Jones index, continually updating its reports on the state of current physical knowledge. This, if it is anybody's job, is the job of physicists. SSK is concerned with the sources and distribution of knowledge in various layers of society at various historical stages. Gottfried and Wilson would seem to be saying something like the following: when Thomas Kuhn set out his brilliant analysis of the gradual revolution that replaced the phlogiston theory with the notion of oxygen, his analysis was flawed because he failed to furnish us with an inventory of all the many new reasons we have for believing in oxygen, as opposed to phlogiston, that have been developed up to the current time. That is, to understand that change, we must also understand almost the whole of subsequent science. They miss the point that it is the mechanism of that change at that time that is the focus of interest, not the physics and chemistry of oxygen [20].

Chapter 13

AWAKENING A SLEEPING GIANT?

Jay A. Labinger

The disagreements that constitute what have come to be known as the science wars have been vigorously argued by combatants from all reaches of a broad spectrum of positions. There does not appear to be much dispute, however, on one point: very few scientists are interested, let alone involved. One of the most prominent of those who are, Paul Gross, is aware of fewer than a dozen scientists active on his side (1998a, 114). If we include all practicing scientists who have participated from *any* perspective the numbers might grow by a fewfold, but not much more. Similarly, Alan Sokal places the main locus of the science wars outside the realm where scientists live: "Finally, *within* academia and the left, this affair tapped into a pre-existing pool of consternation and resentment among non-postmodernist academics in the humanities and social sciences (of which I, as a scientist, was largely unaware). It's this latter factor that has kept the affair going—in the form of innumerable forums, colloquia and debates—in academia."[1]

As Sokal's comment suggests, the disconnect goes well beyond nonparticipation, to nonawareness, on the part of most scientists. Indeed, when scientists I meet ask me what I'm working on, if I mention "the science wars" the reaction is almost invariably incomprehension. This is despite the fact that the events and issues at stake *do* receive at least occasional airing in the "trade" journals that most scientists look at regularly (e.g., *Science, Nature, Physics Today, Chemical and Engineering News*), not to mention their considerable visibility in popular organs such as

1. Alan D. Sokal, posted on an e-mail discussion list, 12 December 1996.

the *New York Times* and *Newsweek,* especially during the months following the eruption of the Sokal Affair in 1996.

It is doubtful whether this situation is alterable to any great extent. Most professional scientists have neither the inclination nor incentives to divert any appreciable fraction of their time and energies from their main pursuits. But should one try? Some want scientists to become more aware of science studies so that they can better defend against the dangers posed by the "antiscience brigades." In contrast, calls for *positive* engagements have been rather rare even on the part of science studies practitioners, who one might think would have a strong interest in enlisting allies from the scientific ranks. I would like to argue that a defensive stance is not warranted and then try to present a case for the potential benefits that could be realized if (at least some) scientists were to take a more serious interest in science studies.

MANNING THE BARRICADES?

A number of writers have proclaimed that science studies pose a threat to science; for example: "Scientists also should confront the sociologists and philosophers at their institutions who are attacking the foundations of science. Presumably, tenure decisions and promotions at universities are based on scholarship, and academic scientists must take an interest in the academic decisions in other departments on campus. This is not a question of academic freedom, but rather one of competency. We should expose political correctness and fundamentalism that lead to misinformation about science" (Bard 1996). This particular statement seems to be imbued with more than a little fundamentalism of its own. Are there in fact inviolable "foundations of science" that may not even be examined without raising issues of hostility or incompetence or both?

The suggestion that science studies, broadly construed, are *deliberately* hostile to the scientific enterprise—consciously aimed at undermining scientific authority and degrading public enthusiasm for science—seems to me plainly unjustified, not least because of explicit statements to the contrary by many of the more prominent practitioners. This point is considered at some length in several of the pieces in this book, so I will not spend any time on it.

However, a claim might be made for *inadvertent* hostility; that is, a claim that this body of work creates a climate in which the abovementioned undermining and degrading goes forward, notwithstanding its authors' intentions: "[T]he fatuities propounded in the name of 'science studies' are not, in themselves, particularly dangerous. I can't envision a host of postmodernists taking to the streets in the name of epistemological relativism. Rather, the long-term danger of the phenomena . . . is that they help erode such defenses against credulity (and they are very meager) as already exist within the culture. In other words, Paul Forman, Sandra Harding, and company, whether they know it or not, are essentially running interference for P. E. Johnson, Duane Gish and company."[2] Such an argument may be defensible on philosophic grounds, I suppose, but in practice? Is there really any likelihood that any significant number among those who believe in creationism (or astrology, or UFOs, or whatever) have the slightest awareness of science studies? Or are more easily influenced by the likes of Johnson and Gish if and when the latter use science studies to support their positions? Or would be the least bit affected if all the scientists in the world spoke out against "harmful" science studies?

A specific arena for concern has been that of science education. Some have argued that increasing influence of science studies will be detrimental (Koertge 1998a); others, that they are *already* substantially responsible for the dismal state of K–12 science education in the United States. In a recent book physicist Alan Cromer places much of the blame for the latter on the prevalence of the constructivist approach to education, which in turn derives much of its appeal, according to him, from the social constructivist turn that science studies has taken over the past couple of decades: "Constructivism is a postmodern antiscience philosophy" (1997, 10). While in no way denying that science education in the United States is in a sorry state, I have trouble seeing how this can be legitimately ascribed to the onset of science studies. The examples Cromer offers of abysmal teaching practice are not an obvious consequence of constructivism: one can teach badly under *any* educational philosophy.

As for his characterization of constructivism as an "antiscience" philosophy, it is intriguing to note his defense of science against a charge of amorality: "The purpose of a scientific study of human behavior isn't to undercut morality, or to dictate conduct, but to enlarge the scope of discourse about such matters. It should offend only those who are completely comfortable with their sect's, or cult's, or party's peculiar view of the world" (79). Suppose we were to change a couple of words? "The purpose of a sociological study of scientific practice isn't to undercut authority, or to dictate conduct, but to enlarge the scope of discourse

2. Norman Levitt, posted on an e-mail discussion list, 8 September 1998.

about such matters. It should offend only those who are completely comfortable with their sect's, or cult's, or party's peculiar view of the world." Perhaps many scientists *would* be offended by the suggestion that they constitute a sect or cult, or that the word "peculiar" (in any sense) is applicable to a scientific worldview. But perhaps many would be sympathetic to the notion that they should evaluate apparently foreign programs with the same degree of charity that they would like outsiders to apply to their own.

None of this is to deny that scientists find profound philosophical differences between the conclusions of science studies and their own opinions, nor that they should strongly argue for their viewpoints—I have done so myself (1995). But if these are philosophical arguments, not wars against the heathen, they need to be carried out in appropriate discourse. The far-too-prevalent practice these days seems to be to simply hold up *apparently* murky or outlandish statements for ridicule—a sort of *res ipso loquitor* argument.

For example, in a recent mail forum on the science wars,[3] one contributor wrote: "Let's see if the magazine will print a particular sentence (a whole sentence!) from the work of a leading postmodernist, Sandra Harding. I claim it gives a much fairer picture of the postmodernist attitude toward, and understanding of, science than anything *Physics Today* has yet been willing to discuss." The sentence, of course, is Harding's (in)famous "Newton's Rape Manual" comment. The writer clearly believes that it is "fair" both to represent a large and diverse group by a single sentence from a single author and to offer that sentence, without any attempt to discern or explain what she might have meant by it, as an overwhelming argument. It's hard to decide which of these attitudes is more disheartening.

Of course, it is easy to see why this is an attractive strategy; no less an authority than Voltaire has extolled its rhetorical potency: "Ridicule overcomes almost anything. It is the most powerful of weapons" (quoted in Hellman 1998, 63). But it is a corrosive weapon, which may as easily be brought to bear against science as for it. That is what concerns me about interventions like the famous Sokal imposture. Originally Sokal's article appealed to me, both for its humorous aspects and as a (self-proclaimed) experimental test of the ability to distinguish whether apparently obscure writing actually conceals deep meaning or is simply obscure. I still believe those are merits. (And, I should note, I don't in-

3. *Physics Today*, January 1999, 15ff.

clude Sokal with those who are content to let "ridiculous" statements stand on their own; in his subsequent writing he has tried to engage with the possible meanings thereof.)

More recently, though, something that happened during the 1998 political campaign has given me second thoughts. A candidate for US Senator repeatedly broadcast ads charging his incumbent opponent with wasting taxpayer money by, among other things, supporting a research project on "cow gas." (The ad was accompanied, naturally, by appropriate background sound effects.) According to some analysts, this and related ads were largely responsible for bringing the challenger from a far-distant second in the early polls to a very narrow loss on election day. But the aspect I found most striking was the fact that the incumbent's campaign managers defended their candidate *only* on procedural grounds: they pointed out that his "support" for the study consisted merely of a vote against cutting a large appropriation for the Environmental Protection Agency, of which the particular project in question was just a small piece. There was no attempt at all (based on the news releases and other pieces posted on the candidate's Web site) to argue that the study in question (which is concerned with the efficiency of cows' transforming caloric intake into milk production as opposed to wasteful—and environmentally harmful—methane formation, and whether improvements might be made) just might be scientifically valid and even potentially important. Once the study had been presented as ridiculous, with no examination of its content whatsoever, any course of action based on actually trying to address the merits of the case was obviously considered much too risky.

I certainly don't want to argue against the use of humor as a rhetorical (or political) device; I would hate to end the science wars by making them so boring that everyone loses interest! But episodes such as the one described above show how the strategy of ridicule can do damage to the overall climate of discourse, and one might hope that it will be wielded judiciously, if at all.

JOINING THE PARTY?

If there is no need for scientists to take a defensive interest in science studies, what is the case for their paying *any* attention? Is benign neglect the right attitude after all? I will argue, for a number of reasons, that it is not. The most straightforward is that some scientists will find science studies interesting intrinsically, just as many scientists are interested in "traditional" history of science. Clearly this alone won't convert anyone who is not of that persuasion, but I believe there are stronger claims.

A number of arenas where the interests and expertises of scientists and science studiers intersect may be readily identified—science policy, political decision making, public understanding of science, to cite a few broad characterizations. Most scientists will readily admit that they cannot claim sole authority over such issues, but they are equally unwilling to cede it to any other group, and mechanisms for integrating the science community with other components of the body politic to reach consensus seem to be lacking. The model most scientists appear implicitly to prefer is: let us do the science, and then the rest of you can make a political judgment based on our firm scientific conclusions.

But surely we don't need the science studiers to point out to us (although they have done so repeatedly) that this model is often (usually? always?) unworkable in practice. For it to succeed, we would need a scientific consensus beyond the point where a respectable set of dissenting scientists can still be found to support an opposing course. Experience tells us this is *very* difficult to achieve. Another problem, perhaps even more serious, is that decisions often must be made before anything close to scientific agreement has been reached.

Global warming may be a good example of the latter. Many experts—probably a majority—are convinced by evidence already in hand, but a very sizable group, including many well-regarded scientists, are not. So shouldn't we take the "obvious" course of holding off on any policy decisions until we have the additional evidence needed to decide the matter? But what if, as many believe, it will then be too late to avert changes of an unacceptable magnitude? If it comes to that, will we then excuse ourselves by saying that yes, in retrospect, things might have worked out better if we had taken action sooner, but as scientists we were acting responsibly by waiting for decisive knowledge? (And will anyone who *isn't* a scientist be much impressed by that defense?)

We urgently need better mechanisms for dealing with such cases, and if scientists are not in a position to provide them on their own, nor willing to hand over authority to other sectors, the only alternative is to foster active collaborations. And that, in turn, requires that scientists begin to look into what is going on in other fields.

The strongest possible reason for scientists to follow science studies, of course, would be to demonstrate that they could have an impact on how one practices one's science. There is no convincing case for this yet; indeed, some science studies practitioners have explicitly denied it. But I believe at least the possibility exists. Scientists have long been dubious whether philosophy of science has anything to offer them, as in Weinberg's quote cited in the introduction, but one of the hallmarks of recent science studies is its emphasis on practice as opposed to theory, as noted, for example, by Pickering: "In the 1970s, academic science studies took a naturalistic turn. Led by a few sociologists, people started examining what scientists did for a living, and discovered that they experiment. It turned out that scientific experiment was not some trivial adjunct to theory" (1998). I must confess that "scientists experiment" doesn't strike me as much of a discovery: I doubt if many chemists, at least, would have ever dreamed of representing experiment as the tail on theory's dog. But there is a valid point here: there is a growing body of work that at least *tries* to be more relevant to the day-to-day practice of science, and scientists who are unaware of it (most?) might well find things there that speak to them.

One frequently voiced objection is that these studies are at best incomplete: by focusing on the social to the exclusion of the natural world they produce severely distorted pictures or even miss the point altogether. I would agree with the part about incompleteness, but how often do we expect a scientific study to be the last word on its subject? In fact, one typical strategy of scientific experimentation is isolation of variables: determining the effect of changing one while holding others constant. We are aware of the limitations of this approach—interactions between variables can well compromise any conclusions we might reach—but we do it anyway, as a useful strategy. The sociologist's practice of "bracketing out" the "correct" results in carrying out his case study can arguably be viewed as a close analog of that practice [14]. I take note of McKinney's objections, addressed specifically to Collins' and Pinch's *The Golem* (1993): "Collins and Pinch and their colleagues have done science studies a great service in choosing to focus on the broader, less methodological issues that naively scientistic accounts of the scientific process have ignored for far too long. But if Collins and Pinch wish to remain true to their title, *The Golem: What Everyone Should Know about Science,* they must tell the whole story, not just the parts that have been neglected. Replacing one incomplete story with another serves nobody's interests (1998, 147)." But this complaint seems to me unreasonable: who can tell a complete story? Isn't adding to, rather than replacing, another story a worthwhile goal?

And furthermore, those (like McKinney) who think science studies *do* have something to offer but have reservations about how they are carried out, may well find that sober and reasoned criticism can stimulate some of the reconsideration they seek. The recent second edition of the aforementioned *The Golem* (Collins and Pinch 1998) includes a substantial afterword, and significant revisions, prompted by scientists' responses to some of the language they found objectionable in the first edition. Perhaps in many other cases, a convergence of views (at least partial) might not be too much to expect!

To illustrate my belief that science studies can offer something to the practicing scientist, here is an example I'm hoping to expand into a full-fledged study: that of so-called bond-stretch isomerism (Parkin 1992). This was a case, first reported in the literature in the early 1970s, where a chemical compound was found to exist in two isomeric forms, differing in color. That in itself is common, but here the detailed molecular structure as revealed by X-ray crystallography showed that the two forms differed *only* in the length of *one* of the bonds. Everything else—all the other bond lengths and angles, as well as the packing of molecules to make up the overall structure of the crystal—was virtually indistinguishable. Such a situation was unprecedented and seemed highly unlikely. Over the next twenty years a number of additional examples were reported, as well as a theoretical study that supported the possibility of that form of isomerism, but in the early '90s a reexamination showed convincingly that the central evidence—the crystallography—was artifactual, a consequence of sample impurity.

What does this have to do with science studies? It is fairly easy to see that the original misinterpretation, and its persistence for two decades, are in large part due to two aspects of X-ray crystallography—both corresponding to concepts that feature in science studies (Latour 1987). First, it has become a highly "black-boxed" technique: much of the collection, analysis, and even interpretation of data is automated. Second, it is a "privileged" technique. In the realm of molecular structure determination, all other methods such as various forms of spectroscopy have become subordinated. (In seminars one almost invariably hears phrases like "to settle the structure once and for all, we turned to X-ray crystallography. . . .") Indeed, one of the more striking features of the evolution of the bond-stretch controversy is that warning signs, coming from spectroscopy and even old-fashioned chemical analysis, that something might be amiss were essentially ignored in the face of the "indisputable" crystallography.

Now, one could certainly claim that this is a fairly trivial application of science studies, and that scientists would be perfectly aware of, and capable of addressing, such issues without any help from outsiders. True enough; but it still seems to me that having them called to one's attention from an outsider's perspective may be particularly effective in raising one's sensitivities and increasing one's alertness to the possibility of analogous situations. Certainly my own initial reaction to the bond-stretch case resolution (before I'd become significantly aware of any science studies literature) was pretty much along the lines of "well, good, they've cleared up a mistake," rather than trying to place it in a much broader (and much more appropriate) context. Would this work in any useful sense for other cases? It would be interesting to try to find out.

We can go beyond such specific cases to larger questions. As many have pointed out, science cannot possibly progress if it spends too much time constantly questioning its fundamental credos,[4] but progress is equally impossible if *no* such revisiting and revising takes place. Scientists seem confident that there is sufficient internally generated reflection, but others are not so sure.

For example, Sokal responded to Bruce Robbins's suggestion that truth can be oppressive as follows: "'It was not so long ago,' Robbins explains, 'that scientists gave their full authority to explanations of why women and African-Americans . . . were inherently inferior.' But *that* isn't truth—it's ideology posing as truth, and objective science demonstrates its falsity. This error is repeated throughout Robbins' essay: he systematically confuses truth with *claims* of truth, fact with *assertions* of fact, and knowledge with *pretensions* to knowledge (1996c)." No doubt the distinctions Sokal draws—between truth and claims, fact and assertions, etc.—are legitimate from a philosophical point of view. But his dismissal of Robbins's argument seems to me *much* too facile. It was "objective" science that was taken to demonstrate, earlier this century, what Sokal now calls ideology. How are we to know, today, whether some of our "truths" are in fact "claims" that might some day be revealed as false? But any such determination would have to be preceded, clearly, by a willingness to suspend belief, and it seems plausible that, again, in some cases an external impetus might be more effective in producing that state. In the case of racial inferiority, the visible example of Nazi Germany may well have done much more to provoke challenges to accepted scientific wisdom than anything initiated by the scientific community itself.

Basically, I'm just suggesting that we need an occasional fresh look from a completely different perspective—even if it looks plainly wrong—to shake us out of too comfortable thinking patterns. In this regard, Sokal and Bricmont's comment about metaphorical aspects seems to me too restrictive: "Some people will no doubt think that we

4. See, for example, Kuhn 1977.

are interpreting these authors too literally and that the passages we quote should be read as metaphors rather than as precise logical arguments. Indeed, in certain cases the 'science' *is* undoubtedly intended metaphorically, but what is the purpose of these metaphors? After all, a metaphor is usually employed to clarify an unfamiliar concept by relating it to a more familiar one, not the reverse" (1998, 10). Certainly that's *one* role of metaphor, but not the only one, maybe not even the most important one. I would focus on the "relating" rather than the "clarifying" aspect: metaphors can forge new connections, and even if most of them lead nowhere, who knows which one will spark a novel, productive insight?

So perhaps science studies—maybe even the "antiscience, postmodern" work assailed by so many critics—is well situated to carry out that function. Even such a nonsympathizer as E. O. Wilson has acknowledged that role: "As today's celebrants of unrestrained Romanticism, the postmodernists enrich culture. They say to the rest of us, Maybe, just maybe, you are wrong. Their ideas are like sparks from fireworks explosions that travel away in all directions. . . . [A] few will endure long enough to cast light in unexpected places. That is one reason to think well of postmodernism. . . . Still another, the one that counts most, is the unyielding critique of traditional scholarship it provides. We will always need postmodernists or their rebellious equivalents. For what better way to strengthen organized knowledge than continually to defend it from hostile forces?"[5]

History tells us that Roman generals, riding their chariots in triumphal parades, were accompanied by slaves who would ride behind and continuously whisper in their ears about the transient nature of the world: what's on top today will be dust tomorrow. History doesn't tell us, so far as I know, whether the generals ever indulged the inevitable urge, whenever a slave got sufficiently irritating, to turn around and whack him in the chops. While in no way pushing the analogy too literally, I would suggest that scientists have been rather quick to respond to voices of skepticism with annoyance. If instead we consider them as invitations to reexamine, even if only occasionally, all that we take for granted, we might well find in them useful spurs to think in new ways.

5. 1998a, 59–62. I should acknowledge that the ellipses in this quote represent much less complimentary comments!

Part Two

COMMENTARIES

Chapter 14

REMARKS ON METHODOLOGICAL RELATIVISM AND "ANTISCIENCE"

Jean Bricmont and Alan Sokal

There seem to be two main themes in the contributions of social scientists to this volume, and we would like to briefly discuss both. First of all, it is claimed that people on the "science side of the science war" (needless to say, we reject such labels, but that is how we are viewed anyway) miss the point in their criticism of SSK: instead of advocating philosophical relativism, their practitioners rely only on methodological relativism [2, 8, 10, 12]. The second point is that it is a mistake for scientists to think that SSK people are antiscience or irrationalists, and here the comments follow two different lines: either it is emphasized that there are far deeper and more dangerous forms of irrationalism in our society, or it is argued that a better (sociological) understanding of how science works will lead the public to a more positive view toward science [2, 5, 13]. Let us discuss these points one by one.

Social scientists who think that we misunderstand them on the issue of relativism actually misunderstand what we say (irrespective of what other "scientific science warriors" may have said). We take pains (both in our contribution to this volume and in our book) to distinguish philosophical relativism from methodological relativism, and we realize that supporters of the Strong Programme often mean to defend only the latter.[1] But our criticism proceeds in two steps: First, we observe that it

1. We also point out, although it is a somewhat secondary issue, that many statements in favor of methodological relativism are so poorly stated (e.g., Latour's Third Rule of Method or the initial formulations of the Strong Programme) that they appear to support philosophical relativism (or radical skepticism) and not merely methodological relativism.

makes no sense for scientists (whether natural or social) to hold philosophical views such as radical skepticism or radical relativism. But then, we argue that methodological relativism is also untenable *unless* one adheres to some kind of radical skepticism.[2]

Since this latter point seems to be remarkably difficult to get across, let us argue it again using an example proposed by Harry Collins in the present volume [12]. Collins illustrates methodological relativism by explaining how a sociologist reacts to the fact that "south of a certain line in a certain country it is generally believed that a certain kind of religious ceremony can turn wine into blood, whereas north of that line it is generally believed that no such thing occurs." He emphasizes that the sociologist, in his explanation, should *not* take into account the matter of whether the wine really does turn into blood [12].

But why is that true? We know that this attitude is common among sociologists, but we nevertheless fail to see any valid reason for holding it. Suppose that the wine really did turn into blood (of course, we know that it does not, but let us just imagine that it does, for the sake of the discussion). And suppose further that people south of the line had discovered solid empirical evidence in support of this (true) claim. In that case, people south of the line would simply be rational in holding this belief.[3] Of course, the sociologist would still have to explain how they discovered the process for transforming wine into blood, and evidently a lot of social factors are at work there. But the explanation would be very different from what it should be if one assumes (as we do) that wine does not in fact turn into blood.[4,5]

2. This point is also well explained in Steve Weinberg's contribution to this volume [9].

3. More precisely, they *could* be rational in holding this belief. It is also possible, of course, that they hold this belief primarily or entirely for reasons of dogma or tradition. Their actual motives for holding the belief need to be investigated empirically.

4. What creates some confusion in this example is that the people north of the line may reject the transubstantiation dogma for reasons that are no more rational than those of the people south of the line who believe in it: they may simply have another religion. In that case, of course, a symmetrical explanation might be justified. (But that too is open to discussion: after all, even among religious beliefs, some are more rational than others. Thus, the people north of the line may reject the transubstantiation dogma simply because they, like Collins and us, are rational people and there is no evidence that wine can turn into blood.)

5. A more sophisticated version of methodological relativism is set forth in this volume by Peter Dear, who begins by observing that "people do not believe propositions to be true *because* those propositions are in fact true" (emphasis his) [10]. Of course, people do not believe statements *solely* because they are true; rather, they believe them (at least in some cases) because of *evidence* that they are true (or at least approximately true). But—and this is the important point glossed over by Dear—the existence of such evidence is often causally linked to the fact that the proposition *is* (approximately) true. Thus, our belief that the Earth is (approximately) round is due in part to the fact that it *is* (approxi-

Sociologists frequently admit that they don't have the background to evaluate whether the claims made by scientists (particularly concerning contemporary research) are rationally justified or not, but then they assert that they are not obliged to make any such evaluation: they are concerned with social phenomena, not with physical or biological ones, and so are perfectly justified in ignoring this latter aspect. That would perhaps be fine if their aims were more modest than those of the Strong Programme: if, for example, they claimed only to recount *some* of the factors affecting the acceptance of scientific beliefs, without purporting to judge their relative importance. But in that case they ought not claim to give a causal account of the acceptance of scientific beliefs, when important parts of the cause—usually the dominant parts, in our view—are excluded a priori from consideration.[6]

Jay Labinger addresses this issue explicitly [13]: "One frequently voiced objection is that these studies are at best incomplete: by focusing on the social to the exclusion of the natural world they produce severely distorted pictures. . . . [But] one typical strategy of scientific experimentation is isolation of variables: determining the effect of changing one while holding others constant." So far so good, but then Labinger goes on to assert that "[t]he sociologist's practice of 'bracketing out' the 'correct' results in carrying out his case study can arguably be viewed as a close analog of that practice." Absolutely wrong! The sociologist's practice corresponds to *ignoring* some relevant variables (i.e., the strength of the empirical evidence), not to holding them constant. One can't hold constant a variable that one refuses to measure [28].[7]

Of course, our objections to the Strong Programme's methodological

---

mately) round: for if it were (for example) flat or tetrahedral, current observation techniques would allow us to know that.

Dear also observes, correctly, that "whatever the evidence and arguments might be, they would count as at least part of the explanation for my belief, regardless of their plausibility to other people." Nevertheless, our evaluation of the plausibility of the arguments *is* relevant to understanding the causal processes leading to the belief: this is apparent from the example of the transubstantiation dogma, as well as from the simpler example ("it is raining today") offered in our own article [3].

6. In passing, let us observe that imposing strict a priori rules on the kind of information allowed to be used in scientific research is radically alien to the methods of the natural sciences. We cannot imagine a chemist or a biologist saying: "Look, no matter what, I am never going to use information coming from physics." He may think that this kind of information is unlikely to be relevant to his work or is going to be too difficult to understand, but these are practical considerations, not matters of principle. Truth is hard enough to find without encumbering oneself with arbitrary methodological limitations.

7. The sociologist's practice is analogous to investigating the effect of religious beliefs on the incidence of lung cancer without bothering to ask whether the subjects are smokers or nonsmokers.

relativism can be circumvented if one asserts that no belief is *ever* objectively more rational than another (and here we mean *really* rational, not merely "rational relative to some socially determined rules").[8] But this then brings us back to radical philosophical skepticism or relativism. That is why we say that methodological relativism is untenable *unless* one adheres to philosophical relativism.

When all is said and done, if the Strong Programme had been called the Weak Programme—emphasizing the limitations intrinsic to the sociological perspective—we would probably never have had any quarrel with it [28].

Turning to the issue of whether SSK is "antiscience" or hurts the public image of science, we'll just make some brief remarks. First of all, we do not judge ideas according to their real or alleged "consequences," nor according to the real or alleged motivations of their advocates or detractors, nor according to the real or alleged moral worth of the social groups in which they are popular or unpopular. Our objection to the Strong Programme is that we think it is philosophically and methodologically misguided, not that it is harmful to the public image of science.[9]

Of course, the reader might wonder why two obscure physicists ever got to worrying about the views held by a fraction of the people working in a subdiscipline of the social sciences called SSK. As we explained in the introduction to our essay, we gradually came into contact with the relativist Zeitgeist that is widespread in certain academic circles, where it is taken for granted that science is merely one "narration" among many others, with no claim to objective validity; indeed, this idea is considered

8. More precisely, we are referring here to epistemic rationality (what one should do if one wants to find out what is true about the world). This is to be distinguished from practical rationality (what one should do in order to achieve some specified practical goal).

9. In reading the contributions to this volume, particularly those of the social scientists, we were struck by the recurrence of emotive words: "undermining" (Pinch), "antiscientific" (Shapin), "threat" and "hostility" (Labinger). Suffice it to say that we have never attacked the alleged motivations of practitioners of SSK, nor have we used the overly vague accusation of being "antiscientific." Rather, we have made precise philosophical and methodological criticisms.

Let us also mention, in passing, that we fail to see what the list of "contentious and provocative metascientific claims" cleverly concocted by Steven Shapin proves [8]. Either it shows that one can misrepresent people's views by quoting an isolated sentence out of context, or that scientists (even famous ones) sometimes express their views in a rather confused way—things we never doubted. But we do not think that we misrepresent the views of a large group of practitioners of SSK when we say that they defend methodological relativism (which is what we criticize them for), and we don't see why confused statements made by physicists or biologists do anything to justify confused statements made by sociologists.

so well established that it need not even be argued anymore.[10] And when we asked for the basis of this remarkable discovery, we were led to what we called the "usual suspects": Kuhn, Feyerabend, Rorty, and the Strong Programme (among many other sources, of course). So we went back and looked at the (alleged) sources. What we found was a mélange: some statements are clearly false, others are ambiguous, still others are reasonable but rather banal; and finally, there is a lot of concrete empirical work, on which we make no judgment [30].

Practitioners of SSK frequently emphasize that they really love science and simply want to give the public a better understanding of how science really works. We applaud that goal, but argue that SSK's sociological reductionism offers a grossly distorted view of the scientific enterprise. Of course, SSK cannot be held responsible for the irrationality prevailing in the general society, which extends from religious fundamentalism to New Age[11] and is obviously far worse than anything we might complain about in SSK. Moreover, SSK's influence probably does not extend much outside academic circles (though it does have some influence on elementary and secondary education). Nevertheless, an intellectual movement that claims to give a causal explanation of the content of scientific theories by treating them, for methodological purposes, just as if they were a religion or a myth cannot be expected to be of much help in fighting the irrationalism prevalent in our societies (to put it mildly) [29].

Our reaction to SSK might be criticized by attributing to us a slightly elitist and old-fashioned view, namely that the universities should be strongholds of scientific rationalism in the middle of the "ocean of insanity upon which the little barque of human reason insecurely floats," as Bertrand Russell once said in another context (1949). But if that is the case, so be it.

10. Of course, it is possible that the people in question are not *really* radical relativists, but simply express themselves unclearly. If so, their ambiguities are quite serious.

11. According to recent polls, 47 percent of Americans believe in the literal validity of the creation account of Genesis, 36 percent in telepathy, 25 percent in astrology, 11 percent in channeling, and 7 percent in the healing power of pyramids. See Sokal and Bricmont 1998, appendix C, note 17, for details and original sources.

Chapter 15

ONE MORE ROUND WITH RELATIVISM

Harry Collins

THE MAJOR THEMES

Epistemological relativism implies that one social group's way of justifying its knowledge is as good as another's and that there is no external vantage point from which to judge between them; all that can be known can be known only from the point of view of one social group or another. Ontological relativism seems to be the view that within social groups such as those described above, reality itself is different. Let us call any combination of epistemological and ontological relativism "philosophical relativism."

Methodological relativism says nothing direct about reality or the justification of knowledge. Methodological relativism is an attitude of mind recommended to the social-scientist investigator: the sociologist or historian should act as though the beliefs about reality of any competing groups being investigated are not caused by the reality itself.[1]

1. My remark to the effect that (we should assume) the natural world does not affect what is believed about it is frequently quoted. It has been set out in different ways in different places. So as to make my own work less confusing to readers, let me put the remark in biographical context. From the early 1970s to about 1980 I thought that my own case studies supported philosophical relativism. After that date I concluded that philosophical relativism was not the sort of thing that could be given empirical support—at least not in the way that scientific theories could be given empirical support. Therefore, philosophical relativism had to be separated from methodological relativism. I set out this view most clearly in a paper published in 1981, in which I said: "[T]he tenet of symmetry implies that we must treat the natural world as though it in no way constrains what is believed to be" (1981b, 218). Throughout that paper I stressed that it was the methodological position which was crucial. Since then, my position regarding philosophical relativism has been that of an agnostic who had found the argument about philosoph-

Philosophical relativism does not make any difference to the practice of the sociology of scientific knowledge. One of the reasons that many practitioners of SSK become frustrated with the continued high salience of discussions of philosophical relativism within the "science wars" and so forth is that they realize that the outcome of the argument does not make much difference to the work being done or to its detailed conclusions.

Why, then, are the majority of the "accusations" in the science wars about philosophical relativism rather than the substance or details of the findings? Part of the explanation is that it is easier to engage in an abstract debate than to get down to cases. But another part of the explanation is probably to do with "the power of words." It seems to be believed by some that the spread of philosophical relativism will damage the prestige of science, making it less attractive to funders or less able to withstand the assaults of creationists, Lysenkoists, and the like; it is believed that epistemological relativism might usher in a new "dark age."[2] Yet none of the relativists I know are remotely in favor of the relaxation of educational standards or the promotion of superstition, nor do I see any evidence for the evil influence of SSK; philosophical justifications tend to follow intellectual and political discontents rather than lead them.[3]

But even if it were the case that relativism is widely associated with dark tendencies, the solution should still not be philosophical censorship. As I argue, and try to exemplify, in my initial contribution to this book, the solution is to make it ever more clear that there is no link between relativism (of any kind) and the value that is placed on science and on what people refer to loosely as "rational thought."[4] The tenability, as opposed to the correctness, of philosophical relativism still seems worth defending against spurious and politically motivated attacks be-

---

ical relativism less and less interesting. One reason it has become less interesting is that I find I would not want to change more than one or two words in the early papers because the change of philosophical position does not make any difference to the case studies; this is the point.

2. For an explicit statement along these lines, see Holton 1993. See also the references cited by Shapin [8, note 3].

3. See also Jay Labinger's remarks on this topic [13].

4. Indeed, Pinch and I argue the contrary in respect of at least methodological relativism in the *Golem* series (Collins and Pinch 1993, 1998b). We argue that to avoid an antiscience reaction it is best to present science as the kind of human activity portrayed in the case studies driven by methodological relativism rather than as an enterprise capable of giving certain answers to difficult questions. The heroic model of science encourages an antiscience reaction when it is compared with the inevitable failures of science in contested technological settings.

cause to do otherwise would be to connive in a witch hunt: imagine that associating with the many worlds theory of quantum physics were suddenly to be seen as a ghastly crime! Philosophical relativism is, then, worth defending because academic freedom is worth defending. In any case, philosophical positions, such as relativism, can be liberating. Philosophical relativism, solipsism, and the like can help distance one from commonsense views of the world. Sociologists and historians need to "alternate" (Berger 1963) between the views of the groups they study, to take them on in the way that an actor has to take on new personalities. And sociologists and historians of science need all the help they can get in the face of the immense consensus-forming power of modern science. Most "SSKers" wish the arguments about philosophical relativism would go away, but so long as they do not, my view is that the academic right to rehearse philosophical positions should be defended. That is all I want to say about philosophical relativism per se.

IS THERE A LINK BETWEEN METHODOLOGICAL RELATIVISM AND PHILOSOPHICAL RELATIVISM?: BRICMONT AND SOKAL

Bricmont and Sokal say that SSK's methodology is a "conceit" and "a diversion from the important matters that motivated science studies in the first place" and that its "practitioners are not obliged to persist in a misguided epistemology; they can give it up and go on to the serious task of studying science" [3]. The odd thing about this conclusion is that they introduce their piece by saying that they are not going to discuss the details of any of the case studies that have been done under the SSK label. Furthermore, they say, "we do not deny that interesting work may have been done in those studies."[5] On the one hand Bricmont and Sokal say that science studies has an important task which it is not doing because it has been diverted by its methodological and epistemological conceits; on the other hand they say that the case studies they criticize may have done valuable work but that they are not going to spend any time looking at them. This is disappointing. It is rather as though they were to say that although quantum theory has come up with useful results, the underlying indeterminacy is impossible to accept, and so they recommend its abandonment forthwith—"quantum theorists should go back to doing sensible classical physics," as they might put it.

5. Adding, gratuitously, "particularly when the authors violated their own declared methodological precepts."

Let us start by countering what seems to be Bricmont and Sokal's central argument—that methodological and philosophical relativism cannot be disentangled [25]. Why bother to counter it given the earlier defense of academic freedom? Because the argument is wrong, and because, regrettably, the argument is likely to reduce the possibility of informed conversation. The defense of methodological relativism and the defense of philosophical relativism for those who care about it are very different in tone and style and should be carried out in separate places. Methodological relativism is a technical matter.

Bricmont and Sokal begin their argument by embarking on the perilous ocean of counterfactuals. Counterfactuals are arguments based on how things would have turned out if we could have altered some aspect of the past. Bricmont and Sokal try to imagine how one would ever explain our society's belief in the inverse-square law of planetary motion as opposed to, say, an inverse-cube law, if no information about planetary movements had been available. They then contrast this with astrology, in which case, they say, the whole set of beliefs could be explained without reference to the movement of the planets. But then they add that if you think that astrology is in fact supported by evidence, then movement of the planets would have to enter into the causal explanation of our astrological beliefs after all. Note again that the initial plausibility of the argument is created by picking long-settled cases where they can expect almost universal consensus about the scientific facts from their audience. In other words, they pick an "easy case" for their point rather than a "hard case." It would have been much more interesting had they picked a hard case where there is a current public controversy [25].

Nevertheless, let us go along with them as far as we can and try to sum up what they are saying: They say that if you think that the natural world affects what you believe about it (that is, if you are not a philosophical relativist), then every discussion of the sources of our beliefs about the natural world must include a natural world "term." Discussions which follow methodological relativism and do not include such a term are, therefore, disguised advocacy for philosophical relativism. Bricmont and Sokal seem to argue, then, for a "dual cause" model: at least in the case of unsettled controversies both social and natural factors enter into the matter and therefore both social and natural factors should enter into all reasonable explanations of the outcomes of such controversies.[6]

Following Bricmont and Sokal we might, then, end up with three kinds of science: type 1, where the current consensus is overwhelmingly

6. The sort of position that Steven Weinberg accepts in his paper [9].

dominated by the natural world—the inverse-square law for gravitational attraction being their paradigm; type 2, where the outcome is overwhelmingly dominated by the social world—astrology being their putative example; and type 3, perhaps live controversies, where there is a mixture of both causes. According to their argument, if I have understood it, even in the case of type 3 science, methodological relativism is untenable because if you know there is some contribution from the natural world you must include it, and that means that methodological relativism, if it is sincerely held, is merely a mask for philosophical relativism. The argument is that if two types of cause arc sincerely believed to be operating in the formation of belief *p* then any sincere explanation of why *p* is believed must invoke both kinds of cause.

We can show that Bricmont and Sokal's argument is incorrect because when it is applied symmetrically it reduces to the absurd. Thus, in type 3 cases there are two types of cause. Therefore in type 3 passages of science, explanatory papers must invoke both kinds of cause if they are sincere and free from error. This means that papers in type 3 areas of science, written by scientists and published in a scientific journal, that did not mention the social factors that contributed to the scientists' beliefs would be flawed. Therefore nearly all scientific papers that have ever been published during the course of any scientific controversy have exhibited insincerity or error. These papers' concern with scientific matters alone must exhibit lack of integrity or an untenable belief in the nonexistence of social factors and effects [25].

There must be as many ways of bringing out this absurdity as there are causes at work in scientific controversies, and there are always many, many, causes at work as any passage of scientific debate moves toward an outcome. One can use Bricmont and Sokal's counterfactual method on any of them. Thus, I cannot imagine the inverse-square law coming to be accepted unless humans had brains of a certain size and unless civilization had not been destroyed by meteor impact prior to the law's discovery. Yet the study of the origins of our belief in the inverse-square law does not have to involve either human anatomy or terrestrial catastrophes.

All this is just a long-winded way of demonstrating what Jay Labinger has said in his paper [13]: a good method in science is to concentrate on only one kind of cause even if others are operating. But it may have been worth going the long way round so as to show what is wrong with arguments based on counterfactuals—they can be made to prove almost anything. Mine is just a version applied to science about the dangers of the use of counterfactuals in any kind of historical explanation. Methodological relativism's recommendation is for the social scientist to concentrate on social causes.[7]

What I have tried to do here is to establish that it is possible to be a methodological relativist without being a philosophical relativist. In spite of what I have said about the tenability of philosophical relativism, all the practically important "scientific" arguments will be carried forward by considering methodological relativism along with the detailed accomplishments that it has achieved and that Bricmont and Sokal have chosen not to examine.

Taking the case studies a little more seriously might also have saved them from the confusion that appears toward the end of their piece where they discuss Hume's argument on miracles [25]. The problem—of what the citizen is to make of competing claims from experts—is one that SSK tries to deal with. Bricmont and Sokal suggest that Hume on miracles has something to teach us: do not believe anything "miraculous" unless you see it with your own eyes. Yet we are nearly all in the position of having the evidence for the "miracles of modern science" reported to us at second and third hand. Bricmont and Sokal themselves discuss the case of a nineteenth-century inhabitant of India confronted with claims about solid water, but the same applies to more or less any modern experiment—for example, all those that invoke quantum theory or the theory of relativity.[8]

Bricmont and Sokal seem to take it that replication of observations is the key. They believe they exemplify the correct approach in their discussion of Jacques Benveniste's claims about memory water. They say "no replication has been claimed, at least not by people totally independent of Benveniste." To make this claim without qualification they must ignore the majority of work that has been done in the last twenty-five years in SSK.[9] While we might agree with them about what the citizen should make of Benveniste, what they provide is the kind of reason that cannot be used without enormous care; it is the kind of reason that can only mislead the citizen and, in the long term, do damage to science by presenting a false picture of it as a quasimechanical problem-solving technique. SSK shows that there are no short cuts, such as invoking

7. For the record, this is not a prescription shared by Bruno Latour and his followers.

8. Science, as well as society, cannot run without trust, as has been thoroughly demonstrated by, for example, Shapin 1994. For a discussion of the citizen's use of the direct evidence of the senses see Collins 1988. For scientists' own worries about the misinterpretation of results by nonmembers of the core-set see Collins 1999.

9. For example, Collins 1992, which is largely devoted to the problem of replication and contains a whole chapter on the notion of the independence of replicators.

Hume's argument, or head-counting, or fraud-busting, or labeling something as "pathological science" to bring the normal processes of consensus formation in science, with all its untidiness, to a more rapid conclusion. Oddly, in this respect SSK puts a higher value on scientific expertise than do Bricmont and Sokal; we prefer the slowly and carefully developed opinions of the experts within the scientific community to any quick fix, such as Hume's argument, that can be applied by an outsider.

Nevertheless, SSK also shows that citizens, when forced to judge between experts, have to make social assessments, not technical decisions.[10] The citizen has to make the same kind of complex assessment as they make when judging politicians: Who has the best track record for integrity? Who has the best credentials? Who is willing to expose themselves to criticism without fear? Who can present themselves with confidence and without prevarication? This point should not be contentious since scientists find themselves making exactly the same kinds of judgments when they encounter findings that they have not generated themselves. In this respect Bricmont and Sokal could learn something from David Mermin's remarks on the social nature of knowledge [7]. Note that only by picking an "easy" case such as homeopathy—where they can rely on the sympathy of their targeted audience—rather than a hard case—such as the health hazards of genetically manipulated foodstuffs—can Bricmont and Sokal's position be made to seem even remotely plausible.[11]

IS METHODOLOGICAL RELATIVISM BORING AND FLAWED?: WEINBERG

The second part of Steve Weinberg's paper [9] seems to be a restatement of the scientist's credo and deals with philosophical relativism. In contrast, the first part touches on issues that directly affect the practice of analysts of science. Weinberg says that historians of science should use our contemporary understanding of science to analyze the formation of the contemporary consensus; I say they should not. Weinberg is saying "avoid methodological relativism—if you know the truth in conse-

10. I am not talking about those with special expertise to contribute.

11. Incidentally, Bricmont and Sokal remark that Collins and Pinch say in *The Golem* (page 144 in the first edition, page 142 in the second) that "if homeopathy cannot be demonstrated experimentally, it is up to scientists, who know the risks of frontier research, to show why." This appears to be a classic (mis)reading out of context. The sentence—and I think any careful reader can see it—is not a defense of homeopathy, but a defense of the role of scientific expertise against the raids of stage magicians and the like.

quence of hindsight, use it." This is a difference, not of credo, but of practice.

As Weinberg himself notes, we have had an e-mail interchange on this topic already. I think some of the points made in the interchange are worth repeating.[12] Weinberg used the example of military history and argued that it is impossible not to be interested in whether some general's assessment of the strength of the enemy, which informed his decision about whether and how to attack, was correct or not—something we do not know until after the event. He argued, I think correctly, that a history of such an event gains in interest from a knowledge of whether the generals of the time had based their decisions on correct or incorrect information. Furthermore, only with hindsight can one ask the interesting question about how they came to have the wrong information, if such it was.

I argued in response that military history and the history of science were different. I said that the relationship between historians of war and generals was different from the relationship between historians of science and scientists. This is because after the event the locus of authority about what really happened in the war—the military fact of the matter—moves from the generals to the historians. (At least, in a democracy, we hope it does, because generals' accounts of the battles they fought are notoriously self-serving.) The historians can open the archives of the enemy army's generals and quartermasters as well as those of the home army. Furthermore, the historian can dispute the causes of victory—and even, on occasion, the military's verdict about who won.

In contrast, the locus of authority on the scientific fact of the matter does not move to the historians of science; it stays with the scientists, so we are stuck with the scientific equivalent of "generals' history."[13] Insofar as they are concerned with facts which are the scientific equivalent of the outcome of battles, historians of science receive them from the scientists. What follows from this, I argued, was that while the military historian had a clear and valuable professional role, it was less clear what the professional role of the historian of science should be. Indeed, the very ground for much of the debate about history of science seemed to be about whether or when the historian of science should treat the history written by "scientific generals" simply as "truth" [33].

Of course, one difference does not necessarily lead to another so now

12. Such exchanges are best conducted on the understanding that they are private unless the authors give permission to be quoted, so I have been circumspect about how I have quoted Weinberg and have made sure he has no objection to the way I have represented him.

13. This is not to say it is always so crudely self-serving.

we need to get into the specifics of the argument. There are lots of kinds of history of science that cannot be done without hindsight. We have tried to classify types of history of science in the afterword to the second edition of *The Golem* (Collins and Pinch 1998). It is only interpretative history which should eschew hindsight and embrace methodological relativism. In this kind of history one tries to put oneself in the shoes of the scientists who were working out what was scientifically true. To recapture that perspective, one must forget one's knowledge of the outcome of that passage of scientific activity. That is, one must ignore what we now count as scientific truth in respect of the passage of science being analyzed. In short, one must be a methodological relativist.

Weinberg says that avoiding hindsight is impossible, or undesirable, or both. It seems to follow that studies of contemporary controversies would be flawed because they do not have the advantage of hindsight. In contrast this seems to me to be their great advantage—one does not have to work so hard to avoid hindsight.

In his attack on methodological relativism, Weinberg first says that it is impossible to reconstruct everything that happened in the past. For example, it is impossible to know why J. J. Thomson favored results at the high end of a certain spectrum of measurements he was making. He says that we know for certain, however, that the correct measurement is the same today as it was in Thomson's time and that the low end contains the correct value. Therefore, he says it is more likely that Thomson was quoting these high values because of prejudice than because he knew he was being more careful when he took the high values.

I do not understand this point. All history (and any contemporary study) suffers from the impossibility of knowing everything, especially in the way of intentions—this is a commonplace. But knowledge of the value Thomson was measuring tells us nothing of Thomson's internal states, since he might have thought he was being careful while he was actually making mistakes of which he was unaware—that is what mistakes are. One cannot read intentions from behavior [33].

Weinberg's second point is that eschewing hindsight is likely to produce boring history because we have to know how things came out to know what is historically interesting. This is undoubtedly true for many types of history of science but stated as a blanket claim it is simply wrong. For example, the history of any contemporary controversy is interesting even though we do not know how it will turn out. And the history of many of those on the losing side of controversies is also interesting. Sales of books about passages of science may go up if the controversy turns out to be important because it founds a new science, but I am sure we are both using "interesting" in another sense—the sense in which the analysis of the first five seconds of telephone conversations or the first babblings of a baby are interesting. That is, we are talking of academically interesting and under this meaning of interesting Weinberg is wrong.

His final point is that methodological relativism "misses the point" of history of science because scientific consensuses are culture-free and permanent. That is what makes history of science different from political history and the rest, which rightly eschew Whiggism. Weinberg says that the influence of culture and the socially consensual nature of scientific laws are only to be seen "while these laws are being discovered." Read one way, this argument is again simply a statement of the physicists' credo, having nothing to do with the methodology of the history of science. But even if the everlasting truth of finally established scientific laws did make a difference to history, it is far from obvious that we should study history only after those everlasting truths have been established. Furthermore, a question with practical import is begged: When do we know that the scientific laws have been finally established? Just when should we take the scientific truths on trust from the scientific generals? For example, how should historians tackle parapsychology? Should we just accept the consensus and run with it, assigning everything the parapsychologists do to one kind of error or another? I cannot imagine a self-respecting professional history that could do this. If one does not like parapsychology, answer the same question for the forty years or so during which the theory of relativity was reaching toward consensus.

I think it is Steve Weinberg who has missed the point of the history of science—at least, the interpretative history of science. The point is that most of the scientific and technical problems we face as citizens concern matters over which there is no consensus—the safety of British beef or genetically manipulated foodstuffs, forensic science as contested in the courtroom, global warming, and so forth.[14] In such cases the political system is asked to make decisions before universal scientific consensus has been reached, yet the model of science that is applied is one where scientists provide the truth for the citizen to use. To understand what contribution science can make to these kinds of problems we need a history of science that looks at science—preferably the best and hardest science—while it is still being made. Perforce, this is the only kind of history we have when we study a live controversy, but if we are inter-

14. For other examples see Collins and Pinch 1998b.

ested in the citizen's understanding of science it is the position we should try to attain when we study a dead one. History of science that works within a world of eternal verities is exactly the wrong model for understanding the way science can contribute to live issues.

OTHER CONTRIBUTIONS

It is an irony of an exercise like this that one has less to say about agreements than disagreements. Wilson and Barsky [11] repeat a common but serious error. Sociologists such as myself are not worried about the mistakes creeping into scientific work or the sudden way scientific disagreement turns to agreement. They are worried that a false picture of science is used to make decisions when science comes into contact with other institutions. A model of science as a fact-generating machine free of human interference is a dangerous model. It is as dangerous to the long-term reputation of science as it is to the citizenry. It also does no favors to the novice scientist.

Both Saulson [6] and Mermin [7] touch on a linguistic issue which may explain the widespread misunderstanding represented by Wilson and Barsky. To the social scientist words such as "negotiation" are neutral. Indeed, their use has been criticized by some social scientists because they are too vacuous. Such terms indicate human interactions whose quality is different to mechanical measurement, calculation, the working through of an algorithm, or the like, all of which are characterized by the determinacy of their outcomes and the inevitability of their course. One may, or I should say must, "negotiate" even when pursuing the truth in the best of faith and even when it is entirely within the confines of what would normally be considered "scientific" activity. Thus, when scientists are breaking out of the experimenter's regress by one means or another, it is not as though they would or could do it better if only they had more scientific integrity (though cases of lack of integrity can still be recognized by other criteria).

David Mermin's first couple of paragraphs show that the representatives of the two cultures are learning to speak to each other. He is right when he says that there is little difference between whether "the construction of scientific knowledge should be viewed as a process of discovering how nature works or as a process of consensus formation among scientists." He is right in saying that the different way of looking at things is really a matter of the questions that are asked and the emphasis that is placed on different aspects of the answer. As for the early history of relativity, it is probably time to leave it to the historians. That there is little between us in the matter of ropes and strands or tapestries and threads is pointed out by Peter Saulson.

My debt to Peter Saulson for teaching me chunks of gravitational wave physics is large. He is one of the people who exemplifies what I most envy about the professional life of physicists—a deep and sincere collective curiosity generally free of the cynicism and pessimism which pervades the culture of academic social science. The natural sciences, when they are practiced with open-minded curiosity, are still the role model for all of us. The one thing that Saulson gets wrong is his claim that a sociologist such as myself necessarily loses interest when scientific success arrives [6]. The large grain of truth in what he says lies in the fact that sociology is much easier when people disagree because disagreement lays so much of the social process out on the surface for easy inspection. But I will be as delighted as anyone if it is agreed in the first few years of this century that gravitational waves have been detected. For one thing, such a success might give my work more salience should it be caught up in the coattails of the new science. More important, however, it will enable me to close my case study by comparing the "same" science under consensus with what has gone before.

Chapter 16

OVERDETERMINATION AND CONTINGENCY

Peter Dear

In some respects, I find David Mermin's essay [7] to be the most interesting of the lot in this collection because of the terms in which he clearly (and fairly) diagnoses the difficulties that had attended his discussions with Collins, Barnes, et al. His "tapestry" metaphor (dimensionally pared down by Collins and Pinch to a "rope" metaphor) lends itself readily to a sharpening of the central issue at stake here: many practitioners of SSK have stressed *contingency* (typically related to Duhem-Quine underdetermination of theories by data) in their accounts of scientific innovation and discovery, arguably at the expense of recognizing that there may, as a matter of empirical (epistemographical) fact, have been a large measure of *overdetermination* in the way in which things came about as they did. The many cooperating threads in Mermin's tapestry are an effective way of conceptualizing this point.

When historians and sociologists of science emphasize the contingency of historical outcomes by making the argument that things could always have come out differently, the point they are making is as much one of methodology as of descriptive accuracy. The methodological importance of stressing a supposed open-endedness in the historical story of science being told is that it counters the Whiggish fallacy to which Weinberg alludes [9]. However, it is worth remembering exactly what that fallacy is. Weinberg represents it as concerning the absoluteness or historical relativity of *theories:* thus, he says that because we can make no real claim to having absolutely superior theories of justice nowadays as compared with those of (say) Martin Luther's world in the sixteenth century, it is obviously improper to give an account of the development of such theories since Luther's time against a background assumption that we now have the right answers. By contrast, he continues, since we *can* claim to know more about the natural world now than people did in the past, there should be no impediment to using that superior knowledge in doing our history of science [33].

Whatever one might think of this claim (and Weinberg's specific examples are interesting in themselves), such an argument does not, it seems to me, go to the heart of Herbert Butterfield's argument against "Whig history." The fallacy inherent in Whiggishness is not one related to some supposed lack of "progress" (of whatever sort) in our beliefs, whether political, religious, or scientific. The Whig fallacy is one concerning historical *explanation.* Butterfield objected to the implicit use of teleological explanations in historical scholarship, and he took as his central examples political histories that used as their "master narrative" an inexorable movement towards a preordained triumph of nineteenth-century British liberalism. In effect, to use such a picture as the backdrop for explanations of particular historical events occurring along the way treats the end of the story as if it had a kind of teleological inevitability, whereby it could be regarded as having somehow *caused* the chains of events that *preceded* it. As such, and as far as it goes, this criticism of Whiggishness appears to retain its validity and applies as much to the history of science as to any other kind of history.

Weinberg states his credo very clearly at the end of his paper, to the effect that he seeks in his work on particle physics some kind of final, absolute truth. As a credo it cannot by definition very well be faulted. It has no value as an argument, however. By contrast, Mermin's concerns have the value of focusing matters on a point lending itself to serious discussion, and that point is a fundamentally historical one. It can be formulated, among many possible forms, in the terms of counterfactual history: to what extent can historical explanations invoke, explicitly or implicitly, notions of what *might have happened*? If Charles Darwin hadn't come up with natural selection as the principal mechanism governing the transmutation of species, would we all still believe in the special creation of fixed species? There are good reasons for supposing that at least some such counterfactuals, although legitimate to ask, should be answered clearly in the negative; taking Wallace among others into account, it is surely indisputable that ideas not just of evolution but even of natural selection itself would have come to prominence without Darwin. Picking up on, and extending, Mermin's "tapestry" image to apply not just to the ongoing composition of technical scientific work itself but also to wider historical processes affecting science, one can translate his argument into a claim of the routine overdetermination of the course of scientific development. If not this experiment in particular, then that and that and that, taken in conjunction, ensure the triumph of relativity; if not Darwin specifically, then Wallace and Huxley and Hooker and (perhaps) Owen together would have ensured the triumph of evolutionary theory.

If this view is embraced, it becomes increasingly difficult to accept the idea that the contingency of particular, individual events is a major factor in producing the outcome of a passage of scientific activity. And yet scholars in science studies routinely use as part of their armamentarium empirically founded claims that the outcome of the scientific dispute (say) that they are studying *could easily have come out differently.* This is a tactic intended to oppose the easy assumption that the outcome of the controversy was really preordained because, after all, nature is really like that. As such, the stress on contingency so common in SSK is entirely appropriate. However, it may be that opposing what might be called "realist determinism" with claims to the radical *social* contingency of scientific outcomes might run into very plausible arguments to the effect that, in whatever scientific controversy is under study, the outcome was historically overdetermined in ways not necessarily related to questions of "how nature really is." That is, even if one wanted, for the sake of argument, to accept a claim that an independently existing nature made no difference whatever to the outcome of some scientific controversy, one might still argue that *social* overdetermination of the outcome swamped any supposed "contingencies" found in the course of the debate. Thus Mermin's "tapestry" image, or Collins and Pinch's "rope" metaphor, would still apply *even if* it were imagined to apply only at the sociocultural level.

If one combines this point with the most common SSK position, oft-asserted (and oft-ignored, as Shapin notes), that the natural world *does* play some kind of causal role, together with other things, in bringing about closure in scientific disputes,[1] then Mermin's point becomes even more reasonable and consonant with much actual work in science studies. Unfortunately, it is a point that clashes directly, as I've mentioned,

1. Collins's position of "methodological relativism" is in effect a bow in the same direction, insofar as it studiously avoids *denying* any role for the natural world; it simply brackets it for the sociologist's purposes. There are arguments, it should be noted, that (for example) Bloor's position on the causal role of the natural world can be criticized as being mere lip-service—as Bruno Latour suggests in his "For David Bloor . . . and Beyond: A Reply to David Bloor's 'Anti-Latour'" (1999a). But such observations take us inside science studies, rather than conforming to caricatures of science studies drawn by unengaged critics [25].

with the "contingency" trick routinely used to counter "realist determinism."

Couched in these terms, one might be tempted to say that the ultimate fault here simply lies (albeit for comprehensible and strategically fruitful reasons) with SSK's obsession with contingency.[2] That would do a disservice to the complexity of the situation, however, for what seems to be a rather subtle reason. In fact, there should be no surprise at its subtlety, because it is a version of one of the oldest and most contentious theological controversies in Christianity: free will versus predestination.

Throughout the history of Christian theology, it has been practically impossible to deny the doctrine of God's omniscience, from which it followed that He knew in advance whose souls would be saved and whose not (in which case, from a God's-eye-view perspective, the answer for each of us had been predestined from the outset). But there has also been the doctrine of the free will of the individual, by which one chooses one's actions, good or evil, and thus deserves one's ultimate fate. When Jean Calvin, in the sixteenth century, stressed the truth of predestination against the emphasis on free will of Roman Catholic doctrine, he was not simply raising a convenient technical objection against the Catholics; he was engaged in an issue of considerable *moral* theological significance. On the one hand, the Catholic church stressed free will because it wished to encourage people to perform good works and wanted Christians to believe that doing so (by their own free choice) would earn them salvation. But by stressing, on the other hand, predestination, Calvin wanted to remove from the souls of believers the burden of worrying about whether their behavior was good *enough* to warrant their salvation: it was, according to his doctrine, a matter of God's will, and what will be, will be. In other words, Calvin's was intended as a doctrine of comfort, even though it looks at first glance as if it ran counter to the Christian duty to do good. Catholic theologians in the sixteenth century themselves acknowledged (it could scarcely have been denied) that predestination was true; God is omniscient. They simply made it a practice not to talk about predestination more than they had to; not to emphasize it to the laity.[3]

The crucial debate between contingency and overdetermination in discussing scientific practice concerns very much the same issues. Mer-

2. See the criticisms of Collins's book *Changing Order* made by Andy Pickering (1987), which determinedly disregard Collins's discussions of core-sets as well as the role of the wider society; see also, for developments of Pickering's position, Pickering 1995.

3. A comprehensive overview of this philosophico-theological debate down to the end of the sixteenth century is Craig 1988.

min's overdetermining tapestry is the equivalent of God's omniscience, while the contingencies stressed in many SSK case studies are the equivalent of free will. Adherents of SSK (such as Collins and Pinch, in the relativity case that Mermin considers) can agree in principle with the general overdetermination thesis (their rope to Mermin's tapestry). But, as with Catholics and free will (despite Collins's claim in the present volume [12] not to understand them!), advocates of SSK prefer to stress, apparently paradoxically, the contingencies that they see as attending scientific outcomes—to show that things could plausibly have turned out differently; that they were not, so to speak, preordained. Hence the point of maintaining that Miller's ether-drift measurements might have been, from a technical standpoint, taken very seriously; and yet, in the main, they were not [30].

Finally, the desirability of emphasizing contingency in the scientific case studies, as with the Catholic wish to play up free will in the theological dispute, stems from a strategic decision. Emphasize free will in relation to salvation and you encourage people to be good; emphasize contingency in relation to methodological relativism and you encourage scholars to investigate scientific practice as a social activity.

The philosophical issues that are involved here, then, are deeply rooted in Western intellectual traditions. As such, it is not remarkable that irresolvable conflicts can arise. David Bloor (1988) has written on parallels between certain responses to the SSK project and responses by some theologians to German Higher Criticism of the Bible in the nineteenth century. The latter case concerned the apparently heretical implications of providing contextualized, historicized explanations of the origins of theological doctrines; critics regarded such explanations as destroying any claim to the doctrines' having a divine, supernatural source. The proponents, however, did not regard their historical work as undermining the theological validity of the doctrines; they just saw it as providing a deeper understanding of how those doctrines had become established. God, after all, was still the ultimate cause of the doctrines' appearance. In the present case, the position of people like Collins is in effect to say that a detailed investigation of the sociohistorical conditions that brought about belief in some scientific truth-claim does not in itself undermine that claim: a Mystery is not required to protect Truth.

Chapter 17

RECLAIMING RESPONSIBILITY

Jane Gregory

Bricmont and Sokal's [3] juxtaposition of Benveniste's controversial work on the memory of water as an experiment in need of replication and homeopathy as a pseudoscience in need of empirical demonstration is interesting because the two topics are intimately connected: the memory of water was seen by some as accounting for the potency of homeopathic remedies.[1] Benveniste is also a surprising choice for those authors to use to extol the virtues of replication, because the rejection of his theory of the memory of water was an exhibition of just how dispensable a tool replication is when it produces inconvenient truths. The experiments, which indicated that water "remembers" solute molecules and carries their activity with it even when these molecules are no longer present, were replicated by several other labs in other countries—indeed, this replication was a condition for publication of the results in *Nature* and accounts for the long list of thirteen authors that heads the published paper.[2] As well as challenging our understanding of many areas of physics and chemistry and potentially making redundant the pharmaceutical industry, the work also offered an explanation for claims by homeopaths that their submolecular solutions could be physiologically active. Despite the positive peer review and the replications that led to publication in *Nature,* a team of investigators led by *Nature*'s editor, John Maddox, set up its own replications in Benveniste's

1. See, for example, Maddox et al. 1988, the *Guardian,* 6 July 1988, and the *Independent,* 27 July 1988.

2. Davenas et al. 1988. The authors are from Toronto, Rehevot, and Milan as well as from Benveniste's lab, INSERM U 200, in Paris.

lab. After seven experiments, *Nature* declared Benveniste's team's extensive data set "an insubstantial basis for the claims made" (Maddox, Randi, Stewart 1988). In its published dismissal of the memory of water, *Nature* reported its dismay that the project had received some financial support from a company that makes homeopathic remedies.[3]

In science, replications, peer review, and publication in *Nature* are usually good enough: the end product is usually well on its way to becoming what Bricmont and Sokal might call "reality" or "truth" [25, 28]. And proponents of the public understanding of science have argued that understanding the peer-review and publications processes is key to the public's ability to distinguish reliable from unreliable knowledge.[4] But what happened to Benveniste would be baffling to anyone who understood those processes as generating truths. Bricmont and Sokal call for replications from "independent" scientists, implying that those who had successfully replicated the work had automatically lost their independence by virtue of their success; the marginalization of Benveniste himself over the last decade is surely deterrent enough to any scientist who might attempt a replication now.[5] Bricmont and Sokal also urge the homeopaths to shoulder the burden of proving their claims, and yet when the homeopaths sponsored Benveniste's reputable, state-run lab to investigate their phenomenon, the fact of that sponsorship was used to discredit his results.

Elsewhere in their chapter Bricmont and Sokal suggest that we need to know how the world really works in order to determine whom we should trust [25]. They are mistaken: if we knew how the world really works, and so knew who was right and who was wrong, we wouldn't need trust at all.[6] Trust is what we mobilize when we don't know who is right and who is wrong—when we don't know how the world really is. In cases such as BSE, when the British public could see perfectly clearly that no one—not the scientists, not the farmers, and not the government—knew what the infected beef might do to the people who ate it,

3. "We were dismayed to learn that the salaries of two of Dr Benveniste's coauthors of the published article are paid for under a contract between INSERM 200 and the French company Boiron et Cie., a supplier of pharmaceuticals and homeopathic medicines, as were our hotel bills" (Maddox, Randi, and Stewart 1988).

4. See, for example, Durant 1993.

5. Benveniste lost his funding, his staff, and eventually his job and has mounted unsuccessful court libel actions against people he feels have accused him of fraud. He is now working in a privately funded lab in Paris. Recently my students Lucy Davidson and Charlotte Burkeman met with some interesting silences when they went in pursuit of Benveniste; he is however vociferous in cyberspace on http://www.digibio.com.

6. See, for example, Giddens 1990.

they turned to whomever they trusted most, often irrespective of expertise, and either stopped or carried on buying and eating beef (Gregory and Miller 1998, 173–80).

I think it's wrong to suppose, as Pinch does [2], that those people who stopped eating beef, or who responded similarly to other scientific "scares," were rejecting science. If two scientists emit opposing pronouncements, however qualified, on the safety or otherwise of beef, they cannot both be correct—at least one of them is either mistaken or lying. When they have a choice, the public do not grant authority to people whom they suspect of incompetence or deceit. This is nothing to do with science in general, nor is it even, in this particular case, to do with the scientific competence or integrity of our two scientists. The reaction is not antiscience but antibullshit: it has a lot to do with the moral conduct we expect of experts, which is judged in the public sphere by the public; and in this regard, scientific experts are no different from any other kind.[7] These moral reactions are of course a problem for science and need to be responded to; but they are not an attack on science itself. Nor are they anything to do with relativism, constructivism, or postmodernism.

Why was it so difficult for the scientists to stand up and admit (which they could have done unanimously) they hadn't a clue about BSE? Pinch argues that the public can cope with expert disagreement, but that they expect certainty of scientists. I suggest that this is a closer description of those who have received a formal scientific education than it is of laypeople; scientists expect that certainty is expected of them, and they expect it of themselves [30]. I disagree with Pinch when he claims that the public are unfamiliar with science being instead as uncertain as everything else in life: the public are not unfamiliar with science being like other things in life, because to the public, science is in life, and it is one of the things in life that things in life are like. Public representations of science-as-certain are pale in comparison with the science of personal experience: so while most people out there in the world have generous, positive attitudes towards some vague but impressive entity called science, they also know that computers can crash and contraceptives sometimes fail; that lead in petrol turned out to be not such a great idea after all and that whether or not red wine is good for you depends on what day of the week it is. The public's paradigmatic science, medicine, is both extremely successful and often haphazard, and its prac-

7. See, for example, Neidhardt 1993.

titioners vary in competence and integrity as widely as do plumbers.[8] Why would anyone who experiences science in these ways look at episodes like the BSE crisis and see in them a reason for rejecting science? Summing reactions such as that against BSE or for astrology into an antiscience movement is something that academics, who like to look for what Saulson calls overarching principles, tend to do, but the public who rejected beef bought microwaveable chicken ready-meals instead, used their cell phones to report the change of menu, and drove to their air-conditioned homes to surf the Net.

In our chapter [5] Steve Miller and I argued that what surveys of public understanding of science most persuasively reveal is the unsurprising news that the public does not think like scientists (Bauer and Schoon 1993). Pinch [2] laments that it is undergraduates rather than the public who are buying *The Golem,* but perhaps it is those who have been educated into the science-as-certainty philosophy in labs and lecture halls who should be reading it; after all, the wider public, who do not think like scientists and who encounter science in life, have already been thumped and coddled by the golem at large. Apart from the fact that no one likes to be told what they "should" know (undergraduates endure it of necessity), perhaps the public are wiser and more experienced judges of experts and their knowledges than we suppose [31].

Pinch argues that it is because we expect science to be perfect that there are antiscience movements in society. There are no antiplumbing movements, he argues, because we've never expected plumbers to be perfect. I'd like to suggest another explanation. When plumbers, lawyers, or nurses let us down, their professional organizations make amends either by dealing with individuals or by making changes to the profession as a whole. All the time, institutions in the social world—presidents, rest homes, bus companies, supermarkets—are reacting and adapting to keep their professional life at peace with society. But science doesn't respond to the public's skepticism or moral censure in the same way: when the public are indignant about BSE or exploding space shuttles or pollution (or don't give a damn about positrons or the orbit of Mercury), scientists insist that the public amend its ideas about science: it is society that has to change to accommodate the scientists, rather than the other way around. Where science does take responsibility for keeping its own house in order, as it did in the case of Benveniste, or of the British scientist who was fired for suggesting that GM foods might

8. On the relationship between the general and the particular in public attitudes to science, see Bauer, Petkova, and Boyadjeva 2000.

be harmful, the resulting spectacle is invariably clumsy and reflects poorly on all concerned. Where once we might have talked about the social responsibility of scientists, now we talk about the public understanding of science [27]. Perhaps this shift of emphasis is a positive step in terms of democracy and the influence of the public sphere, but surely the one need not be a substitute for the other. If there are antiscience (but not antiplumbing) movements in society, then perhaps it is because science does not reflect too often or too deeply on itself.

COMMENT FROM STEVE MILLER

In championing his "expertise" model of science I am concerned that Trevor Pinch [2] may end up in the same position as those who took a constructivist standpoint on SSK to mean that science was "nothing but" a social construct. While I accept that for the purposes of sociological research, or for the purposes of the public understanding of science, the expertise model provides a standpoint from which to carry out research or to work out how to treat the information that has been provided, I feel that science is an outcome of and, now, the driving force of, a rationalist worldview of a law-governed universe. Science goes beyond expertise in that it provides a way of thinking about and investigating the natural world that can be applied in almost all circumstances, familiar and novel. There may be, for example, a cooking worldview; indeed, Heather Couper, a well-known popularizer of astronomy in the UK, once likened the Big Bang to baking a cake. But I do not believe that the cooking worldview is ever seriously applied to cosmology or the theory of evolution, for instance, in the way that physics, chemistry, and genetics are applied to the study of cooking and the modification of the food we eat.

Chapter 18

SPLIT PERSONALITIES, OR THE SCIENCE WARS WITHIN

Jay A. Labinger

A number of contributors to this volume have suggested that what we have here is a failure to communicate and have offered a number of correctives. Most of these can be broadly characterized as: try to avoid words that might be interpreted as something not really meant; try to avoid reading messages that aren't really there. (I was struck by the number of contributors who remarked on the tendency of readers to interpolate the phrase "nothing but" or the equivalent into passages where it may never have been intended.) Among the many themes of discord and concord that have been generated, I would like to concentrate on two that appear, occasionally, to reflect some degree of *internal* division, where individual commentators seem to be making use of both sides of an issue.

I emphasize the words "appear" and "seem" in the above: few if any of the issues we deal with are so black-and-white that one could or should line up unambiguously on one side or the other, and that's certainly the case with the two dichotomies in question. But even the perception that the author is trying to have things both ways, I believe, can be a significant contributor to the misunderstandings and tendencies to talk past one another that have been so well documented in the first part of this book.

First: Is science very much like common sense, or very different from common sense? Entire books could be (and have been—see Wolpert 1992; Cromer 1993) written on this theme, and several of our contributors, including Shapin [8] and Lynch [4], have addressed it, representing it primarily in terms of different opinions held by different scientists.

The aspect I am specifically concerned with is: to what degree is one justified in extrapolating from commonsense or "easy" cases to claims made with respect to science in general? For example, Bricmont and Sokal [3] use the "obvious" question of whether or not it is raining as a starting point to address the role of reality in determining belief. They do recognize that there is an issue here: one can question whether "ordinary" and "scientific" knowledge should be treated the same. Yes, they conclude: if reality constrains ordinary knowledge, it constrains scientific knowledge even more so because experimental science is carried out expressly to make that the case. Thus science would seem to be just an elaborate form of common sense.

On the other hand, Bricmont and Sokal acknowledge that science is *difficult.* Hence it is far from clear that intent—the basis of the claim above—can guarantee outcome or that commonsense arguments can be freely carried over to the consideration of cutting-edge science. It may be straightforward to go from observation to belief in the case of rain, but it is considerably less so in the cases examined by science studies, which are aimed precisely at whether experimental activity is capable of making "Nature itself constrain our beliefs" as Bricmont and Sokal claim. So the appeals to common sense and science may well belong to separate realms, and an argument that seems to freely hop from one to the other is not very convincing.

A particularly problematic connection between the familiar and the not-so-familiar is the analogy between science and the law. Bricmont and Sokal suggest that the statement "it is true that X is guilty . . . but . . . this truth came into being as a result of decisions about how we should licence our police investigations . . . a truth brought about by agreement to agree" sounds very odd. Really? If I substitute "O. J. Simpson is not guilty" for "X is guilty" in the above, the statement sounds like a pretty accurate one to me!

I expect Bricmont and Sokal would counter this by contesting my usage of the word "true" here. Their intuitive notion of "truth" presumably has little to do with a formal jury verdict. But if we have no access to *that* kind of truth, as must frequently be the case in the realm of criminal investigations, arguments based upon it are not particularly helpful; and science studies would say, I presume, that applies in the realm of scientific investigations as well. Here again Bricmont and Sokal appear to be alternating modes of argumentation, starting from a nonrigorous, commonsense understanding of "truth" and proceeding to a closely argued position on whether methodological relativism logically entails philosophical relativism. This gives the nonsympathetic reader an easy opening to dismiss the argument and avoid more productive engagement.

The second issue, stated generally, is: Are the foundations of science studies primarily theoretical or empirical? Obviously this is a huge topic, one that far exceeds my space and resources to deal with adequately. I focus specifically on one aspect that science studies seem to appeal to frequently, either explicitly or implicitly—the principle of underdetermination, or the Duhem-Quine thesis. Both Mermin [7] and Bricmont and Sokal [3] characterize this idea—that there are always any number of ways to interpret any finite set of observations—as logically correct but practically irrelevant. I would have to agree, in the sense that no scientific case history that I know of (even as represented in science studies) reveals any such multiplicity of interpretation after a substantial amount of evidence (a vague term, admittedly) has been collected. What such studies do show (perhaps "claim to show" would be more appropriate, depending on who is reading them) is that consensus can often run ahead of what the data would seem to compel.

But that falls well short of justifying, from an empirical standpoint, any sort of radical underdetermination thesis. If anything, *over*determination is the more common situation: we can find cases where at some point in time *no* theory is found to be compatible with all existing data by the community.[1] That directly contradicts assertions like the following (not from this volume) and makes them rather hard to take: "It is unproblematic that scientists produce accounts of the world that they find comprehensible: given their cultural resources, only singular incompetence could have prevented [high-energy physicists from] producing an understandable version of reality at any point in their history."[2] Such sweeping claims tend to undermine credibility for those readers who perceive therein a hefty degree of prior ideological commitment. Science studiers proclaim the *empirical* nature of their programs, but their empirical work appears to be informed by philosophical tenets that are unsupported—or even contradicted—by observation. It becomes all too easy to dismiss such studies, even when solid data have been collected.

Unlike Bricmont and Sokal, who call upon science studiers to abandon their "misguided epistemology" [3], I wouldn't deny anyone the right to a theoretical framework. But I do feel that a little more care over

1. See, for example, the account of the "missing" solar neutrinos in *The Golem* (Collins and Pinch 1993, chap. 7).

2. Andrew Pickering, as cited (and criticized!) by Galison (1987, 10).

how much weight theoretical grounds are given might help avoid some of the problems of perception. I note Shapin's comment about a connection between weak methodological discipline and success in the natural sciences [8] and wonder whether it might be usefully (and symmetrically!) applied to science studies.

One thing a reader can do is dissect out and discard any theoretical component of such a study and concentrate on the residue, which will often be a valuable, if mainly descriptive, account. For example, whether or not one agrees with Perutz's reaction to the Geison book discussed by Dear [10], it would be hard to argue that recovering the contents of Pasteur's (highly illegible) notebooks was not a heroic effort. Likewise, Saulson [6] characterizes Collins's collection of aural archives as a great service, whatever one thinks of his conclusions.

I recognize that the above may well sound at best like damning with faint praise; it is reminiscent of the "sandwich theory" of reading that Pinnick offers (and rejects) with respect to a particular work: "[The book] just needs to be read with charity. You take the first and last chapters with a large degree of intellectual tolerance (those are the parts to ignore, with all the bad argumentation and the trumpeting about sociology of knowledge); then, in between, there is a good historical account to read" (1998, 228). I wholeheartedly support the idea of "reading with charity," but I don't think it implies any of Pinnick's evident condescension. In many of the scientific papers in my field that both report and interpret data, I find the data extremely useful even though I consider the interpretation to fall somewhere in the range of unconvincing to utter nonsense. But that doesn't mean the endeavor wasn't worthwhile.

Even an effort based on an unconvincing premise can make a valuable contribution. Lynch comments [4] on the potential value of pushing apparently far-fetched ideas, such as Fish's baseball analogy, as far as one can: we can learn from where and how it breaks down. Physicist-turned-biologist Max Delbrück said something similar. In trying to apply Bohr's notions of complementarity to biology, he considered the conceptual transfer as just another research tool. He wanted to push it as far as it would go, recognizing that it would surely break down—but the point where it did break down would teach us something fundamental (Roll-Hansen 2000). Perhaps, as Collins's focus on "strangeness" [12] seems to suggest, this sort of "defamiliarization" is the most useful contribution that science studies can make.

Chapter 19

SITUATED KNOWLEDGE AND COMMON ENEMIES: THERAPY FOR THE SCIENCE WARS

Michael Lynch

In the large body of writings on the science wars (I hesitate to call it a literature), it is common to read historical, social, and personal explanations of the conflict. Constructionists are accused of being '60s refugees who have displaced frustrated political hopes and turned their rage on fellow academics. "Scientists" and the "defenders of science and reason" are accused of being frustrated by recent cuts in basic research budgets, so that they turn their rage upon the "academic left" rather than the fiscal conservatives who made the cuts. Such explanations are variants of ad hominem argument, in as much as they do not engage the interlocutor's position on its own terms but instead try to undermine it by reference to its personal or cultural origins. I think, however, it is worth dwelling for a moment on arguments to the effect that scientists frustrated by the dwindling status and resources in, say, particle physics are scapegoating "sociologists." A university at which I worked for six years—Brunel University, in West London—was named after Isambard Kingdom Brunel, the great nineteenth-century engineer. A few years ago, the Brunel administration closed down its departments of chemistry and physics. I found this news astonishing at the time. From experience in the United States before moving to England, I was accustomed to a situation in which administrators in a budget-cutting mood would turn first to the "soft" targets in the humanities and social sciences. So, why would a university eliminate the two "hardest" fields of the hard sciences? To my knowledge, there was no serious suspicion at Brunel University about the possibility that science and technology studies (which are prominent at that university) were at fault for encouraging administrative hostility toward physics and chemistry. Instead, the decision was governed by a highly *rational* attitude, albeit a narrow one. The university's leaders employed what they called a "model" for weighing the costs and benefits of each member of faculty and each departmental unit. According to this model, the income from teaching and research in chemistry and physics did not cover the costs of running those departments, so, reluctantly, the administration decided to eliminate the departments. (Permanent staff were reallocated to other units.) According to the same model, other areas like sports science and business studies were thriving.

I mention this incident not in order to suggest that members of embattled fields and departments in the physical sciences are jealously projecting hostility upon their colleagues in the humanities and social sciences. Instead, I mean to suggest that the science wars may, in part, be symptomatic of a malaise that afflicts all sides of that alleged conflict. Preoccupations with science and scientific rationality can distract us from recognizing the common conditions in academic life enjoyed and suffered by members of science, humanities, and social science faculties. Whether identified as natural or social scientists, or simply scholars, we lecture to students, attend faculty and committee meetings, publish our research in academic journals, and attend professional conferences. We also have common complaints about the incessant evaluations, audits, and reviews conducted by university administrations and other education authorities. Hope for peace can be found in the recognition of a common way of life, and particularly in the identification of common threats to that way of life. Individual administrators are not the main source of the problem. (Although, as I know all too well from spending several years at Boston University under John Silber's tyrannical presidency, individual administrators can be a disaster for morale in out-of-favor departments.) The problem is what is sometimes called the "business model" of a university and its attendant rationality.

One of the prominent tendencies in constructionist studies of science is a resolute insistence that science is *work*. Like other forms of work, scientific practice is viewed as an embodied and material labor process involving numerous, often obscure, parties (including what Steven Shapin [1989] calls "invisible technicians"). Connected with this emphasis is a stress upon the "local" or "situated" character of scientific and technical work (Suchman 1987); work that involves practical actions and reasoned judgments which are not a matter of mechanically following methodological rules (Collins 1992). Moments of creative struggle and opportunities for improvisation in a laboratory occur from top to bottom in a hierarchy of research directors, staff scientists, technicians, and civilian participants. The contingent products of this collective labor process (data, results, publications, discovery claims) are more than deliberately planned outcomes, as they can be sources of surprise and puzzlement.

An understanding of the implications of this picture of scientific work may provide a basis for solidarity rather than epistemological infighting. Not too many years ago, the autonomy of science was upheld as a generalized ideal, which was necessary for scientific progress. The "new" post-Kuhnian sociology of science systematically attacked this and related ideals as an ideology which was based on rather misty and flimsy notions about Science (notions not even so solid as to be built on sand). Criticisms of the idealized view of science inspired many historical and ethnographic studies, which documented how science is, and always was, embodied, organizationally embedded, supported by patronage, and answerable to political agendas. The concerted effort to demolish the "myth" or "ideology" of scientific autonomy strikes some commentators as a wrongheaded and cynical exercise, but what is often missed about the emphasis on situated practices in science is that it implies more intimate, but no less necessary, modes of autonomy than were promoted by the grand narrative of scientific progress. If, as is so often claimed in constructionist studies, scientific practice is not thoroughly governed or determined by overarching norms of rationality, then it is a mistake to think that top-down administrative efforts to manage and assess scientific "productivity" will have the desired consequences. I do not have the space here to pursue the political implications of this picture of practical autonomy, but I am convinced that they confound many "radical" as well as "conservative" political agendas. At the very least, they provide a possible point of leverage for resisting the models and schemes of assessment that promise to rationalize contemporary academic "business."

What I have said thus far has little direct bearing on the specific contents of the debates associated with the science wars or the more peaceful arguments represented in this volume. In my article in this volume, I stressed that the debates associated with the science wars do not pit scientists against antiscientists, and I also noted that the polemics in these debates should not be mistaken for scientific arguments. Instead, the debates are philosophical in scope and content, even though they are not always (or even often) voiced by professional philosophers. I used a rather irreverent baseball analogy of playing sandlot philosophy with pickup teams.[1] One may want to ask, what is the point of playing such a game, especially when it may seem to distract us from our "real" research? Our insight into this question can be deepened by reading a recently published book on social constructionism—Ian Hacking's *The Social Construction of What?* (1999)—written by one of the most accomplished and influential philosophers of science today. Other philosophers have written about and contributed to the science wars, but Hacking's book represents the best attempt thus far to *take seriously,* in an admirably detached and insightful way, both sides of this often-polarized argument. As he makes abundantly clear, his sympathies for the several species of constructionism are, at best, mixed, but he clearly distinguishes constructionism from an antiscientific attack on the natural sciences. Hacking drives home the lesson that "the science wars are founded upon, among other things of a more political or social nature, profound and ancient philosophical disputes." But, instead of pursuing a peaceful resolution, he identifies "sticking points . . . philosophical barriers, real issues on which clear and honorable thinkers may eternally disagree" (68). So, then, why bother with such arguments if there is so little prospect of agreement?

Perhaps we can explore this question by examining the ongoing exchange between David Mermin on the "science" side, and Harry Collins and Trevor Pinch on the "sociology" side regarding the question of the experimental support for relativity theory in the early-twentieth-century physics community (see Mermin's discussion in this volume [7] for a blow-by-blow account of this exchange). It has been conspicuously nonwarlike, as both parties have clarified and qualified their positions in response to questions and criticisms by their interlocutor. So, in this sense, they pursue understanding and agreement. But, in light of Hacking's analysis, it is pertinent to ask if agreement on more fundamental matters can, or even should be, expected. Hacking would probably counsel against any such expectation. The argument between Collins and Pinch and Mermin is highly focused on the response by physicists to Dayton Miller's 1933 experiments. Collins and Pinch (1993) argue that by 1933 the "culture of life in the physics community" was committed to the theory of relativity, so that the rejection of Miller's results was not based on a definitive refutation, but rather on the presumption that

1. I should mention that when I was younger I loved playing pick-up baseball and am not at all averse to playing sandlot philosophy. I should also mention in this aside that I am stunned by Jay Labinger's mention [13] of how very few natural scientists actually have taken part in the science war games. I believe Labinger, but it is startling to consider that so many articles and e-mail messages have been written by so few.

the results *could not be correct.* Mermin objects that in the years between 1905 and 1933, physics had amassed an entire "tapestry" of evidence in support of relativity and that the "culture" of acceptance should not be reduced to an incorrigible belief. As Mermin recounts, Collins and Pinch then respond "by emphasizing that much of the other evidence available at the time was also far from clear cut" and that "strands of evidence can be woven in different ways." At this point, we might begin to get the weary feeling that this debate can go on indefinitely [30]. Now, as I understand it, the lesson from Hacking is not that we should despair about the unlikely prospect of agreement. This metaphysical argument has been going on for millennia, though of course earlier versions of it did not concern debates about the role of experiments in the acceptance of relativity. Mermin expresses an insight along these lines when, after reflecting on the exchange with Collins and Pinch, he says: "I'm increasingly persuaded that any issue one can formulate in one language has a parallel formulation in the other."

It isn't enough to say, however, that what we have here is a metaphysical disagreement. The content of the disagreement is about a particular sequence of events involving a relatively small research community. The form of argument is historical, even if one suspects that incorrigible assumptions about historical contingency versus inevitability may animate it. Just as experimental physicists sometimes argue by challenging the adequacy of apparatus and technique and by countering one set of experiments with another, historians argue by arranging archival citations and quotations into significantly different narrative sequences. Part of the drama is to see what the interlocutors will do next: give ground, give up, walk away in disgust, or come up with some new evidence. To say that the parties are reiterating metaphysical positions says little about the eventual terminus of their dispute, because their moves in the game must be made and countered with documentary evidence, the furnishing of which takes work and can be a source of surprise, consternation, and even knowledge.

But what is the point of coming to terms with the arguments if we figure that there will be no final vindication, refutation, or dismissal? One point has to do with recognition: a form of déjà vu, "here we go again." This recognition awakens us to the fact that we are taking part in a recurrent drama involving intellectual ancestors and contemporaries. Hacking points out that science wars arguments about the stability of Maxwell's equations and the second law of thermodynamics often move quickly, almost imperceptibly, from accounts of specific experiments, laws, and matters of fact to broader arguments about truth, the nature of facts, and the independence of nature. These more general arguments, to borrow Hacking's down-to-earth characterization, are about the uses and implications of "elevator words" like "truth," "reality," and "fact," and not about particular facts, entities, and equations (Hacking 1999, 21–23). This might suggest that whenever we recognize that we have been pounding metaphysical tables and kicking metaphysical rocks we should feel silly, but such a reaction would trivialize traditions of debate which have had a very long run in the history of ideas. Even if we figure that there is not, and never can be, a scientific solution to these debates, it does not mean that we can ignore them. To recognize the historical background and familiar form of our arguments can be therapeutic in the sense that we now are able to realize that what we thought was a novel and provocative adventure has been following ancient argumentative pathways. Although we may have believed that further pursuit of an empirical research program or a rational examination of the evidence in hand would guarantee the truth of our argument, we now can recognize that each (seemingly fresh and novel) point we make has a well-established counterpoint.

I am not recommending a metaphysical reduction of the "nothing new under the sun" variety, because prospects for a particular argument can be highly uncertain and the details of specific cases can be crucial. As in law, traditional principles and precedents do not determine how a singular case proceeds. However, when in the heat of argument we remind ourselves that we are engaged in metaphysical debate, this can have therapeutic consequences.[2] In the context of the science wars, such reminders may not provide a simple remedy. The therapeutic effect is likely to have more to do with the ethics of argumentation than the outcome of any argument. To anticipate and acknowledge that the interlocutor can keep coming back with reasons and evidence can take some of the steam out of the self-righteous outrage that fuels the science wars.

2. The analogy with therapy, and the role of the philosopher in "assembling reminders," which dissolve metaphysical problems, comes from Wittgenstein (1953, sect. 127).

Chapter 20

REAL ESSENCES AND HUMAN EXPERIENCE

N. David Mermin

No reliable body of knowledge can be undermined by viewing its acquisition as a collective human activity. Indeed I would go beyond Harry Collins and maintain that scientists can benefit from "estranging themselves from their own practices" in this way. What else was Einstein's recognition of the conventional character of simultaneity? Or Steven Weinberg's warning against "ill-placed loyalty" to "the great heroic ideas of the past"? I suspect that many of the interpretive confusions plaguing the foundations of quantum mechanics reflect insufficient estrangement from unsound linguistic practices.

I understood from the start that setting aside the "physical truth of the matter" is a *methodological* principle for sociologists. My "complaint" was not that they do it, but that by doing it too rigidly they can blind themselves to important aspects of an episode of scientific practice. This is what Weinberg [9] and I [7] suggest in our remarks on "Whig history." Trevor Pinch's observation [2] that it is circular to explain the emergence of truth by reference to its truthfulness (as Jean Bricmont and Alan Sokal so often do [3, 25]) trivializes the real issue: whether our current understanding of the truth is so dangerous a contaminant that its judicious use must be categorically forbidden to the carefully trained, suitably forewarned, philosophically sophisticated historical investigator. Peter Dear's remark [10] that "modern scientific beliefs are in fact irrelevant to an epistemographical account" can be too limiting when those beliefs can give clues about the objective circumstances of historic events [27].

Harry Collins [12] misreads Kurt Gottfried and Ken Wilson when he takes them absurdly to insist that SSK should continually update all its reports to incorporate current understanding. It is not silly to insist that a discussion of our understanding of relativity as of 1933 ought to convey an adequate indication of the evidence available up to 1933. It's fine to stop the clock too soon to incorporate all of contemporary understanding provided you're not making a point about the character of contemporary knowledge.

Michael Lynch [4] puts his finger on what I find so irritating about many texts on both sides when he characterizes the "science wars" as a series of exercises in "pop metaphysics" or, even better, as "sandlot philosophy with pick-up teams." In defense of this casually sophomoric tossing around of Great Ideas, one can note that even professional philosophers have not been notably successful in articulating the character of scientific knowledge, so why not open the game to all and sundry? But at least the professionals are aware of the complexity and historical depth of the issues and respect the need for caution and precision.

How any practitioner of physics in the twentieth century can fail to appreciate that "the relationship between language and the world" is highly problematic is beyond me. When Lynch asks [4] whether physical laws might be less like rocks in the field and more like "rules governing human actions," he evokes Bohr's "[I]n our description of nature the purpose is not to disclose the real essence of the phenomena but only to track down, so far as it is possible, relations between the manifold aspects of our experience" (1934, 18). There are important differences between these formulations, but sorting them out is an instructive and nontrivial undertaking—not a matter of delivering one-liners from the sandlot. The same can be said about what distinguishes the five "irritating" assertions, exhibited by Jean Bricmont and Alan Sokal [3], from Steven Shapin's list [8] of eleven "provocative metascientific claims" made by pillars of the scientific establishment.

Shapin's admirable essay misses, however, the point of Mara Beller's piece in *Physics Today* (1998). Beller is not urging a more thoughtful attitude on physicists by pointing out that the wisdom of Bohr would sound like nonsense if it came from sociology or cultural studies. Quite the opposite. She is denouncing the great icons of quantum physics for uttering what she takes to be nonsense, and she is urging scientists to clean up their own act before they get on with the business of mocking others.

This is a manifestation of one of the less noted ironies of the science wars: the presence, on the physics team, of a deviant subculture of "Bohmians" who take the conventional wisdom on the nature of quantum physics to be not only socially constructed, but fundamentally unsound. Examples can be found in Sheldon Goldstein's article in Gross, Levitt, and Lewis 1996 (119–25) and in Gross and Levitt's bizarre denunciation of Stanley Aronowitz because "he naively echoes . . . the view that the causal and deterministic view of things implicit in classical physics has been irrevocably banished" (1994, 52).

Poor Aronowitz subsequently changed his tune to incorporate the Bohmianism of Gross and Levitt (Aronowitz 1996, 181), only to be denounced from the majority side by Kurt Gottfried: "I do not know any theoretical physicist who is thinking about transcending the wave-particle duality . . . for the simple reason that we do not have a shred of empirical evidence against this, or any other aspect of quantum mechanics" (1997).

My colleagues Jean Bricmont and Alan Sokal clearly delight in sandlot philosophy. Since anyone can join a sandlot game, I can't resist noting that the view "that there is nothing to objectivity except intersubjectivity" starts from the fact that we each have nothing but our own sense impressions from which to construct the world but then notes, interestingly, that when you and I learn how to communicate, we discover that certain features of our separate subjective experiences are shared. From such coherences in our private subjectivities we are able to infer that which we call objective. The fact that "once upon a time, people agreed that the Earth was flat . . . , and we now know that they were wrong" does not establish that there is more to objectivity than intersubjectivity. All it shows is that in this instance the inference of the objective from the intersubjective was inadequately based or badly drawn [25].

In the revered words of Wolfgang Pauli, such sandlot argumentation from either side "isn't even wrong." It is banal and unpersuasive. Nothing is accomplished by arguing about whether the truth of a proposition reflects the way the world is or whether it reflects the way people have agreed to talk about the way the world is, as two and a half millennia of philosophy have surely demonstrated. These are roughly isomorphic languages for describing the contents of knowledge. Even when intelligent life is discovered elsewhere in the universe and is found to share our understanding of the periodic table of chemical elements, we can argue about whether this establishes the objective character of the elements or demonstrates important interplanetary transcultural features of how social organizations apprehend inanimate matter.

This does not mean that there is no content to disagreements between scientists and sociologists. But to get to serious issues one should focus on Steven Shapin's "something . . . that is demonstrably wrong, as judged by the consensus of expert practitioners" [8]. When I criticize Collins and Pinch for suggesting that the only grounds for believing in relativity in 1933 were the Michelson-Morley experiment and the solar eclipse expeditions, it doesn't matter whether I'm criticizing their description of the social basis for physicists' beliefs or their failure to record fundamental features of physical reality already discovered by 1933. When Barnes, Bloor, and Henry assert as a textbook illustration of the malleability of categories that "it is not a relationship of similarity or resemblance which permits us to decide what the next case of carbon will be," and I complain that they never mention atomic structure in their survey of the various properties that might be used to define carbon, my objection remains the same whether it addresses their incomplete characterization of the microscopic structure of the inanimate world or their ignoring an important aspect of how chemists go about their business.

Another hallmark of the science wars is exaggeration. Bricmont and Sokal [3] attribute to Barnes, Bloor, and Henry (BBH) "a rather tolerant (or even favorable) attitude toward" astrology and cite my remarks about BBH's use of astrology as offered to refute BBH's favorable claims. But all I was saying is that astrology is too extreme a case for a textbook example of an area currently outside science that might conceivably shift back in. My "Humean type of reasoning" was not to refute an argument for astrology, but to expand on Bloor's remark that their point was well illustrated by my own unwillingness to take off a year or two attempting to duplicate the data claimed by Gauquelin. To be sure, Barnes read me as saying about BBH and astrology what Bricmont and Sokal actually do say, thereby demonstrating yet again that misreading, oversimplification, and exaggeration are not peculiar to any one of our cultures [25].

While I agree with Steven Weinberg [9] that "to decide to ignore present scientific knowledge is often to throw away a valuable historical tool," I'm less enthusiastic about his claim that a scientific theory "is culture-free and permanent." This is closer to the truth than the sandlot Kuhnian view that now and then absolutely everything comes up for grabs and we can start all over again with something completely different. But "permanent" is too strong a work for what Weinberg actually describes. It undervalues what he acknowledges to be changes in "our understanding of why the theories are true and also our understanding of their scope of validity." I'd be surprised and disappointed if particle theory ever reached his "fixed point." Surprised because we seemed to be arriving there too many times before; disappointed because if we actually did arrive, then, as Richard Feynman famously remarked, philosophers would be able to explain why it could never have been otherwise, and nobody would be able to refute them.

On the other hand I entirely agree that the asymptotic character of the process is insufficiently emphasized by SSK. Whether you want to characterize this as "a cumulative approach to truth" or an evolutionary process "driven from behind" seems to me a matter of taste, unless you share Weinberg's faith in the existence of a final theory. Ken Wilson and Constance Barsky [11] beautifully illustrate the asymptotic character of scientific knowledge and the challenging opportunities it offers for new kinds of sociological analysis by their example of the development over the centuries of greater and greater precision in astronomical calculations and measurements. Students of the scientific process seem insufficiently interested in the fact that Newtonian mechanics is alive and well in spite of all the revolutions of the past century, and that within it "normal science" has made steady, remarkable, indeed, "revolutionary" advances of its own.

I wish I'd driven fifty miles up Interstate 81 to meet Peter Saulson before I wrote my first column on *The Golem*. It would have accelerated the process of convergence of opinion I reported here. He and I initially perceived the same shortcoming in SSK views of science—what originally struck Saulson [6] as Collins's "silence concerning the nature of the intellectual debates that accompany the social process of crystallization of a scientific conclusion." What we both came to realize—he, sooner than I—was that much of the disagreement is over the choice of language. A lot of SSK would indeed make more sense "if the discussions of 'negotiations' among 'interested' parties . . . were clarified so that the primarily *intellectual* nature of the interests involved were made plainer." I'd even drop "primarily"; in the present state of SSK that kind of interest seems much too readily overlooked entirely.

I share the skepticism of Jane Gregory and Steve Miller [5] that science studies and "postmodernism" are significantly undermining public understanding of science. Extraordinary claims are being made to this effect. Paul Gross believes generations of undergraduates have been deluded into thinking they understand the physics of relativity by an obscure commentary by Bruno Latour on an early Einstein popularization (Gross 1998b). I have heard the miserable state of science teaching in the California schools blamed on the rise of postmodernism in the academy and wondered why it was already so terrible when I was in high school in Connecticut in the early 1950s.

Finally, to return to Harry Collins, while "guarding truths" against informed critical reexamination can cheerfully be condemned from all sides, much of the heat of the science wars has to do with the legitimate guarding of a body of knowledge from misinterpretations and misreadings which both social and natural scientists should surely strive to avoid.

Chapter 21

IT'S A CONVERSATION!

Trevor Pinch

The overwhelming impression from reading through the contributions to this volume is not only the temperate tone but also the growing sense that we are in a real conversation. There is a well-known finding in the field of conversation analysis that participants in conversations display a proclivity to end in agreement. The sense of moving toward agreement which one finds in the first round of contributions is surely further indication that a real conversation is taking place.

That a conversation is a social event should not be forgotten. It is surely no accident that the productive exchange (remarked upon by several contributors) between David Mermin and Collins and Pinch has been accompanied by many face-to-face meetings. Mermin himself notes the importance of his having spent time with David Bloor for his forming a charitable attitude toward Bloor's intentions [7]. He contrasts this with the less productive discussion he has had with Barry Barnes, whom he has never met. Peter Saulson's obvious respect for Collins as a social scientist has also been shaped by his growing personal ties with Collins [6]. As Steve Shapin notes in his piece [8], most participants in the "science wars" share far more than they disagree over. Ironically the public debate which the science wars has occasioned can serve to remind us of that. Alan Sokal and I recently took part in a radio debate, and I was delighted at one point to hear him defend me against a mediator who simply failed to understand the intellectual issues at stake.

I start with these observations not to suggest that the solution to the science wars is to replace vigorous debate with the establishment of a cozy club where scientists and their newfound friends in science and technology studies can bond with each other, but to point to the obvious sociological truth that fear and loathing are more easily turned into "relativistic cleansing" when you never have occasion to sup with or break bread with the opposition. The strange thing about the science wars (as Michael Lynch [4] and Jay Labinger [13] note) are the odd bedfellows who sometimes find themselves associated with one position or the other. What is hopefully clear to all now is that David Mermin does not hold the same position as Alan Sokal any more than does Harry Collins hold the same position as Sandra Harding.

I must admit to finding the format that Labinger and Collins have devised for us to continue this particular conversation peculiarly difficult to manage. It is as if I have been invited to a cocktail party with many fascinating guests with whom I can only spend a limited amount of time. Do I, like most academics at cocktail parties, stand in the corner glad to talk in depth to the one or two persons I know well already—say debate with my friend and colleague Peter Dear whether epistemography really does capture what science and technology studies is all about [27]? Or do I immediately collar a new guest, say Steven Weinberg, and spend all my time trying to inform him of how we see things? Or do I spend all my time in another corner chatting with David Mermin, picking up on previous conversations, trying to clear up the last residual points of disagreement? Or should I act like the perfect cocktail party guest and have quick conversations with as many people as possible? In the end I decided to do what a good host should do—try to talk to the strangers in an effort to make them feel at home. At the risk of offending Peter Saulson, David Mermin, and Jay Labinger, not to mention all my colleagues in science and technology studies who are gathered at this particular party, I've ended up talking mainly to Steven Weinberg, and Kenneth Wilson and Constance Barsky. I also have a few words for the noisiest guests at the party, Jean Bricmont and Alan Sokal.

I was pleased to hear that Steven Weinberg no longer finds what he calls the "methodological antirealism" position of the Strong Programme absurd, although I sense that his "just a matter of a few minor things wrong with it" means more than he implies [9]. One thing that it is important to get straight, as David Mermin and others have been at pains to stress, is the question of language. The correct term for the position under discussion is not "methodological antirealism" but "methodological relativism"—a term which these days Jean Bricmont and Alan Sokal have learned to use [25]. To take but one example: Barry Barnes has gone on record explicitly arguing that his position is realist. Barnes rejects what he calls "double-barreled" realism, which posits correspondence between scientific entities and what is out there in the world, but nevertheless his benign realist position should be correctly acknowledged as such. The point is made excellently by Dear. There is nothing inconsistent about believing in reality and doing the epistemological bracketing on methodological grounds (epistemography in Dear's terms) associated with most work in science and technology studies [10].

My major disagreement with Weinberg—although I am tempted in like spirit to call it "minor"—is on the matter of how much present knowledge should be allowed to influence the retelling of episodes in the history of science. Weinberg is surely right to note that we cannot know everything in a historical reconstruction [9]. But the danger of allowing present knowledge to play trumps seems to me evident from his own example concerning Thomson. Weinberg wants to use our present knowledge of the ratio of mass to charge of the electron to suggest that some of Thomson's measurements were distorted by his commitment to a favored value which he developed early in his measurements—a value which was actually too low according to current knowledge. I, like Weinberg, do not know if Thomson was actually more or less careful during his measurements, but it seems dangerous for historians to impute care or sloppiness based on what we now take to be the "correct value" since this begs the question of the pristine nature of the correct value. Philip Mirowski (1994) has shown in a very interesting article how measurements of physical constants tend to fall around what is taken to be the "best value" for a time until a new "best value" comes along (known as the "bandwagon effect"). If we apply Weinberg's position on this matter to these other cases, it would seem that we are saying that virtually all work for long periods of time in measuring important numbers is sloppy until finally we have a careful measurement. With lack of direct evidence of sloppiness, I would rather be charitable to Thomson and other experimenters and say let us at least look for other reasons as to why they might have favored a particular value [33].

Mirowski's work is also of direct relevance to Wilson and Barsky's call for research on the long-term sociological factors that affect scientific results [11]. Mirowski argues that the "bandwagon effect" means that "the uniform approach of physical constants to some invariant asymptote is itself a myth" (1994, 579). The problems are well known amongst metrologists, and this has led to new and more elaborate institutions and social structures in science for the allocation of quantitative error, such as meta-analysis. In short what Wilson and Barsky call for is already being looked at in one part of science studies [34].

Another issue raised by Weinberg is the danger that history of science without hindsight will become boring. Of course, what makes a subject boring or interesting depends on what perspective you approach it from. Conversation analysts find the first few seconds of conversation interesting because they are looking for the systematic properties of turn-taking—on the other hand most historians would find the first few seconds of no interest whatsoever. Often historians of science do use hindsight to pick the most salient cases to examine. Indeed, in my own work on the history of the solar neutrino problem (Pinch 1986), I argued that we should pick cases where scientific controversy is already turning to consensus. What needs to be distinguished are two different uses of present knowledge. There is little difficulty in using present knowledge to find interesting problems to work on, and indeed it is scarcely possible to do historical research at all without this level of knowledge (this is a point also made by David Mermin), but this is not the same as using current knowledge to judge the scientific competence of historical actors at the time.

Another matter worthy of comment is the place of Thomas Kuhn's work in science and technology studies. It may surprise our scientist commentators, but the reappraisal they call for [9, 11] has already happened, and Kuhn no longer has the influence he once had. Although *The Structure of Scientific Revolutions* continues to be a ritual citation, his specific ideas are not today part of an ongoing research program in science studies. There is a consensus that his notion of paradigm was too broad brush for studying everyday science, although it has proved useful in some studies of cataclysmic change in science (e.g., Collins and Pinch 1982). Kuhn's own subdivision of the paradigm notion in subsequent writings has proven to be too vague to be empirically useful, although paradigm as "exemplar" continues to be a very influential metaphor. Indeed, modern science studies, as Dear notes, has turned away from the epistemological question of Does science progress? to much more answerable questions concerning how specific bits of knowledge and specific institutional arrangements in science and technology get to be the way they are. The grand epistemological issues have by and large been put aside. It is rather like physicists working in quantum mechanics putting aside the unanswerable question of whether matter is made of waves or particles. Some areas advance best by deciding which questions it is meaningful to ask.

One issue which also still seems to cause trouble is that of pseudoscience. Bricmont and Sokal are disturbed by a reference which Collins and I made to homeopathy and suggest that SSK's methodological relativism leads to a "rather tolerant (or even favorable) attitude toward the pseudosciences" [3]. Mermin still seems exercised by Barnes, Bloor, and Henry daring to address the matter of astrology in the same vein as they treat other more acceptable areas of science [7, 30]. It is true that within science studies there has been a focus on "pseudo" and "fringe" sciences. The reason for this has not been because of any particular fascination with such areas. The reason to look at pseudoscience is because looking at the abnormal tells you something about the normal. It is a similar strategy taken by physicists who look at the unusual or extreme case as a test of their ideas. No comfort for pseudoscientists should be drawn from such work, as Jay Labinger rightly notes [13]. I suspect some scientists here find it hard to bracket their own concerns about pseudoscience. From our perspective, where science is *our object of study,* pseudoscience looks like a particularly interesting boundary case.

With these general comments in mind I turn to the specific allegation made by Bricmont and Sokal. When we wrote "if homeopathy cannot be demonstrated experimentally, it is up to scientists, who know the risks of frontier research, to show why," we were not recommending a shift in the burden of proof away from the advocates of homeopathy. The subject matter of the quoted passage was not homeopathy per se but what we saw as the illicit use of simpleminded views of science to police the work of scientists who have come up with experimental evidence for unorthodox theories. We wanted to draw attention to, for instance, the heavy-handed debunking operation which *Nature* coordinated against one such scientist, Jacques Benveniste, who had published his findings therein. Our suggestion was that this debunking operation was no substitute for the normal means of scientific peer review and replication of experiments. There is nothing particularly radical about our attitude toward what befell Benveniste—many scientists at the time were equally dismayed by *Nature*'s treatment of him. Our wider sociological point was that we wanted to draw attention to the passing of responsibility in judging such matters from the scientific community to another community, where stage magicians and journalists hold court.

Last, I welcome the comments made by Labinger, Wilson and Barsky, and Saulson that the sociology of science should focus on the interaction between science and the wider society. Much cutting-edge work in the field indeed does just that. The piece by Jane Gregory and Steve Miller [5] nicely illustrates some of the current themes, but their chapter is only the tip of an iceberg of a whole range of work orientated toward politics, communication, the law and policy (e.g., Jasanoff, Wynne, Lewenstein, Ezrahi). Indeed the *Golem* series was written precisely because we wanted to make a contribution to that domain. In the UK today, media pundits, politicians, and scientists alike bemoan the policy disaster which the issue of genetically modified food has become. Yet it is precisely here where science studies expertise is of most use in counteracting overly simplistic versions of what science can and can't deliver. The sad thing about the science wars is that they threaten to distract from this work and belittle the expertise of those who do it.

Chapter 22

CONFESSIONS OF A BELIEVER

Peter R. Saulson

I believe in the Church of Baseball.
—Annie Savoy, in Ron Shelton's *Bull Durham*

I believe in the Church of Science. Harry Collins's analogy between the sociological study of science and the sociological study of religion is more apt than he let on. We scientists fit the profile of true believers who are convinced we have privileged access to the Truth, and who are confused about and suspicious of people who want to treat our belief system as a social phenomenon [25].

Most of my fellow scientists are also clearly true believers. Says Steven Weinberg, "I can't *prove* that the laws of physics in their mature form are culture-free . . . [but] I am convinced that it is so" [9]. His faith is so strong that he offers as a fact a *future* miracle, in the form of a final theory of physics, in order to show that Science is the Way, the Light, and the Truth.

The analogy is useful because it illustrates clearly why we scientists, for all of our on-the-ground experience in the living activity of science, should not be taken by ourselves or others as the best, let alone the only, experts on how science works. Outside observers with a background in understanding social life clearly have a primary role to play in elucidating the function of science as a social system, embedded in a larger society. Natural scientists who think otherwise are indeed guilty, to use the term of several contributors to this volume, of practicing sociology without a license.

Recognition of this state of affairs would be valuable for everyone interested in science and how it works. The fact that many scientists are prepared to dispute it is a measure of how far the social study of science has to go to gain legitimacy [25]. That front of the science wars is and should be an active one; it appears to be a major feature in the discussion in this present volume.

There is another useful analogy in *The Golem at Large,* Collins and Trevor Pinch's companion book on technology (1998b) to *The Golem* (1993). One of their case studies involves the monitoring in Cumbria of the nuclear fallout from the Chernobyl disaster. The essay focuses on the relationship between the expertise of nuclear scientists and that of sheep farmers much more intimately familiar with the land and its life. The lesson Collins and Pinch draw is that the episode would have been handled much more successfully if the outside experts had had the proper respect for the local detailed knowledge of people without formal credentials.

The relevance of this analogy for the science wars is that sociologists play the role of the outside nuclear experts, while the natural scientists are the sheep farmers. For all of their acknowledged expertise, the outsiders need to take proper account of the kind of knowledge that the locals have, and even to pay proper respect to what the outsiders might consider simply local prejudices. (Of course, good sociologists of science already do this.)

The rest of my remarks in this essay should be taken as those of a sheep farmer to whom experts should attend in spite of my lack of learning, or of a believer who is well-placed to witness a faith that is of interest to others.

WHENCE COMES THE ANGER?

The science wars represent a new phase of the problem of the two cultures. When Snow wrote, the biggest division between humanists and scientists was a simple ignorance and lack of interest in the other side's concerns, which resulted in at worst a passive mutual contempt. Now we have a situation where several camps within the humanities and social sciences have the natural sciences as the focus of their activity, some with an explicitly critical agenda. A formerly demilitarized zone has been crossed. Few of us on either side are good at the fighting that has resulted.

Some in this volume question whether war is a good metaphor for the dispute between scientists and practitioners of science studies. Of course, in many respects it is not. But it has to be admitted that there is a remarkable degree of hostility involved in the exchange. In my role as a witness from the science side, I would like to try to explain the origin of the anger that scientists feel. This may help to put into context some of the less admirable aspects of the science wars.

Do people really care enough about metaphysics and epistemology to make them angry? Aside from a few philosophers, my impression is that most don't. In particular, few scientists of my acquaintance pay any attention at all to the philosophical issues that surround science.

Scientists do care about politics, some of us passionately. Alan Sokal explained in his *Lingua Franca* article (1996a) that his intention in perpetrating his hoax was to help reclaim the left for the forces of reason from the mystics who hold it hostage today. Few other science warriors are so explicitly leftist; indeed a more common attitude would be that science studies, postmodernism, and what passes for the left on campus are all one big mess. Even so, few scientists care strongly enough about their politics to explain the anger behind our actions in the science wars.

To explain what I take to be the source of the anger, let me report my experiences upon first hearing about Sokal's hoax article in *Social Text.* I happened to be away from home, serving on a proposal review committee for the National Science Foundation. The news reached me by e-mail, in a message from a colleague to the rest of our department. I don't think the glee that I read in the message was injected by me.

Several of the other members of the review committee had known Sokal when he was an undergraduate. (He and I were classmates.) Since I knew this, I shared the news of the hoax during a break in our meeting. There was much amusement, some pride at his success in pulling off the stunt, and a discussion of whether the person we had known years ago was the sort we should have expected to do something like this. None of us expressed any concerns that perhaps there was something impolite, uncollegial, or less than scholarly about his action. I believe that most of us felt that he had "scored one for our side."

How was it that we already felt enough disdain for our nonscientist colleagues on campus that we instinctively rallied to Sokal's side?

We take pride in our feeling that, as scientists, we are engaged in an activity that is distinctively different from that of our nonscientist colleagues. In a very real sense, we think it is better. Science progresses. Problems get solved. Knowledge accumulates. Old disputes are put to rest. Weinberg, in his essay in this volume [9], testifies vividly to this belief.

The corollary to this belief is that we think that most of the work of humanists or social scientists does not share these traits. We see age-old debates that never end. Worse, there is a lack of agreement even on fundamental questions and the ways to answer them. As a result, disciplines are divided into factions that can't even speak to one another (sometimes only metaphorically, but sometimes literally).

Ordinarily, a veneer of collegial respect prevents scientists from telling their nonscientist colleagues how dismaying their activities appear to us. "Transgressing the Boundaries" transgressed this boundary of civility. Sokal gave our hidden feelings a voice and gave us a vicarious thrill in the process. For a moment, it was as if we were claiming the privileged status that we feel our group deserves, as successful generators of knowledge in an academic culture where knowledge is the highest ideal.

The fields labeled "science studies" consist of a group of our nonscientist colleagues who, instead of ignoring us, look back at us without appearing to share our own valuation of our collective worth. This is in itself enough to provoke feelings of suspicion and mistrust. Worse than that, though, sociologists of science appear to us not even to be doing their own job right. We see science as an institution that works uniquely well. Those who study us instead see an activity much more like other human activities than it is distinctive.

This disagreement is not (just) over something as abstract as epistemology or metaphysics, nor as vaporous as politics. It goes to the heart of the reasons most scientists feel good about their work, and hence about themselves. We know that, individually, we are no better human beings than anyone else. But we do take immense pride in being participants in such a successful and worthwhile activity.

Why is the view from science studies so different from our own collective self-portrait? Why is not the distinctiveness of science at the center of modern treatments of it? In other words, why can't they get it right? Two possibilities that might spring to mind are incompetence or malice. Either would be worth fighting, whether to expose the first or to combat the second. In either case, a righteous anger is a useful spur to carry on the struggle.

I hasten to add that I subscribe to neither explanation offered above for the disjunction between scientists' self-image and the portrait painted by sociologists. By having followed Shapin's prescription [8] of discussing these matters in cafés and pubs (mainly with Harry Collins), I have learned that science is in fact a much less distinctive activity than it first appears. Deep in my heart, I confess that I still hold the faith that something about it is both unique and uniquely good. But identifying and explaining that feature also seems to be uniquely hard. My own guess about the reason for the failure of sociologists to "get it right" is that they are confronted with a truly difficult problem. In such a circumstance, there is no shame in being stuck. Sociologists of science are doing just what scientists do when confronted with a problem that is too hard to solve, turning their attention to other problems that are solvable.

Perhaps I have confessed too much of my own inner life in these paragraphs, and quite possibly I have attributed to others feelings which are mine alone. But unless we recognize such feelings where they exist, we will continue to trip over them.

THE BATTLE OVER METHODOLOGICAL RELATIVISM

Much of the discussion in this volume turns on the issue of whether methodological relativism is simply a methodology of sociologists who study science, or whether it embodies a conclusion about the nature of scientific activity itself, specifically that sociological factors rather than scientific factors govern the progress of science. Jay Labinger's description of it as the equivalent of a scientist's controlled isolation of variables [13] may be easier for scientists to hear than the equally clear but less familiar sounding defenses of it by Collins [12] and Dear [10].

The essays by Weinberg [9] and by Bricmont and Sokal [3] both attack methodological relativism as misguided. The treatment of the latter two authors disappointed me, since they descended to an almost Scholastic exercise, turning their backs on the credo of science to respect how the world really is and instead arguing their case from first principles. Their piece was built around a close reading of the manifesto of the Strong Programme of the sociology of scientific knowledge, written by David Bloor. Bricmont and Sokal tried to show that, by its very nature, SSK is built upon the premise that sociological factors determine the actions and beliefs of scientists [25].

The flaw in Bricmont and Sokal's textual analysis is easy to spot. Remarkably, they build their case using only on Bloor's principles 2 and 3, while completely ignoring the explicit statement in principle 1 that "Naturally there will be other types of causes apart from social ones which will cooperate in bringing about belief." That statement, grudging as it may sound, is clearly intended as recognition that scientific progress can happen for good scientific reasons.

To be fair, Bricmont and Sokal have probably not had much of an opportunity to gather empirical facts about the practice of the sociology of science by adherents of SSK. I have, at least to the extent of having had numerous long discussions (and some arguments) with Harry Collins during his ongoing study of the field of gravitational wave detection in which I work and of having read a reasonable number of his papers and books. So I can pretend to be something of an amateur sociologist of the sociology of science. Based on my empirical observations, I can assure my fellow scientists that Collins's application of methodological relativism is much as Labinger describes it. If his is supposed to be the face of the Devil, we should all lighten up.

Chapter 23

BARBARIANS AT WHICH GATES?

Steven Shapin

Supposing the science wars ended and the science warriors won.[1] What kind of culture would the victors then inhabit?

Here is one scenario: sociologists of science are held up to ridicule and contempt by their academic colleagues; fingers are pointed in senior common rooms and faculty clubs; tenure and promotion to the higher rungs of their career ladders are denied; their students cannot find suitable academic employment; universities conduct a disciplinary cleansing and, ultimately, sociology of science disappears from the academic scene.

No big deal, it might plausibly be said, even by those who do not actually intend these results. We're talking about maybe a hundred men and women whose careers are at stake (possibly more, probably less); the suffering involved is scarcely on the scale of Rwanda or Kosovo; and the economy will chug on without their contributions. It's always good for academics—structurally prone to portentousness—to remember in such circumstances A. J. Balfour's dictum: "Nothing matters very much, and most things don't matter at all." I think this is a thing that *does* matter, just a bit—I would, wouldn't I?—but I do not fool myself that many other people think so. Strange as it may seem to some of us caught up in the conflict, I have to report my personal experience that almost none of my scientist-friends have even heard of the science wars, and,

1. The opposite outcome is only logically possible; indeed, every sociologist I know wants nothing more than to decline combat on the terms offered.

when I tell my nonacademic friends about it, they shake their heads in bemused disbelief.

On the other hand, those victorious science warriors who talk about "hating" members of the so-called Edinburgh school would take pleasure in work well done.[2] The supposed science-haters would have been liquidated and the culture made safe for genuine science. Yet, on reflection, such a victory might prove Pyrrhic. It would have been achieved over the systematic protestations of fellow academics that their views have in fact been misrepresented, and a precedent will thereby be set whose working-out in university culture should give everyone pause: members of one discipline will be licensed not just to evaluate the work of another in which they are not trained, but they will be authorized to characterize its points and motives in ways that members of the indicted discipline systematically reject [32]. When sociologists say they have no reason to dispute the present-day laws of physics, they will be disbelieved; when they say they do not hate science but, rather, that they see their inquiries as part of science, they will be told that others know better what motivates them; when they insist that their object is a naturalistic understanding of science, they will be informed that they dissimulate.

I will not use this occasion further to protest my own innocence of any such crimes, or the innocence of that of my sociologist-colleagues in this volume: I merely wish to draw attention here to the structural implications of a situation in which a group of academics are obliged repeatedly to attest, not just their competence in their special subjects (which, though unpleasant and uncommon, is certainly fair enough) but the innocuousness of their states of mind and intentions in doing their specialized work (which, in my opinion, is not). More generally, the cultural phenomenon of academic "antiscience" would be accorded a legitimacy it does not deserve, namely, positing its substantial and co-

2. No one in this volume, of course, has been intemperate enough to say such things in print, though, unfortunately, I have heard many such professions, as well as endorsements of the intended outcomes, expressed viva voce over the past several years. However, should it be thought I am making things up, see the sentiments of the English biologist Lewis Wolpert, quoted in Roger Highfield, "Science and Sociology Fight for Grip on Reality," *Daily Telegraph* (London), 11 April 1997: "I attack them [the Edinburgh school] at every opportunity, I hate them. They are the true enemies of science. These people are the kiss of death. They have a political agenda to control science themselves and to diminish it at every possible step." As a former member of the Science Studies Unit at Edinburgh University, I must minimally record my incredulity and deep disappointment at such remarks: no such intention or "political agenda" was ever part of my consciousness, at Edinburgh, or before my time there, or since my departure; nor was I ever aware of such intentions or agendas on the parts of my colleagues at Edinburgh. And it is difficult to imagine what possible "agenda" could allow three low-paid British academics, even if they were so minded, to "control science."

herent existence. The dismal fate of the academic "antiscientists" would then be available as an object-lesson of what risks attend certain forms of inquiry. The university will be seen as a haven for free inquiry just on the condition that you naturalistically study subjects less vocal and less powerful than yourself. The lesson will be: study down; do not study up.

But if you do not think that academic antiscience is either substantial or coherent, still less consequential—and my major contribution to this volume gives some reasons that I do not—then the science warriors are wasting a lot of very valuable time and energy. Science will be the poorer for the waste of all that distinguished time and energy. One is inevitably reminded here of Monty Python's wonderful *Life of Brian,* where the Judean Liberation Front pours its passions into fierce ideological conflict with the Liberation Front of Judea: it's an easier target than the Romans, and small differences among in-groups are always more exciting than the large ones that divide the in-groups from the external hordes. Or, if you prefer the version commonly attributed to Henry Kissinger, the reasons academic disputes are so violent is that so little is actually at stake.

Were the health or public credibility of science really threatened by the writings of a few evil-minded sociologists of science, then much would be at stake. If there is any genuine expertise in how academic ideas travel and how they affect public attitudes, then it probably resides in the social sciences. Yet I am unaware of any plausible social scientific scheme, or empirical study, that shows an effective and linear causal relationship between specialized monographs and the sentiments of millions of people who provide no evidence that they are even aware of such monographs.[3] By contrast, there is much respectable sociological and historical work establishing the responsiveness of academic views to widely distributed cultural sentiments.

And, as is common when sectarians concentrate on purifying the academic groves, real and substantial problems outside the groves go neglected, or, worse, the delusion spreads that sectarian purification rituals are effective ways of addressing such substantial problems. Some science warriors are worried about "public misunderstanding of science," and, in my view, rightly so. The proportions of Americans who profess belief in astrology and in the separate creation of species—figures, by the way, generated by sociologists—are really appalling. Anyone who refers

3. The relative sales of, say, Gross and Levitt's *Higher Superstition* and Bloor's *Knowledge and Social Imagery* would be pertinent. I do not know these figures, but whoever shows me that the former is not much larger than the latter gets a bottle of the excellent 1995 Château Cantemerle.

blandly to the Victorian "triumph" of science over religion or superstition cannot be living anywhere near my neighborhood.[4] Nor should one assume that the problem is one pertaining just to the Great Unwashed. At my own reasonably distinguished university, I have taught many biology students who are creationists and some astronomy students who believe in astrology, and I can assure you that they were not taught these things, or encouraged to believe these things, by sociologists.

Jean Bricmont and Alan Sokal attribute the credibility of flat-Earth or witchcraft beliefs to "the existence of a radically relativist academic Zeitgeist" [3, 25]. Again, I am not sure how they have established this interesting causal hypothesis, but matters may not be so simple as they seem. How would they propose to explain the heterodox beliefs of my science students (and, for all I know, the beliefs of some of their own students)? We're too rigorous—may I say "scientific"?—in my bit of sociology to allow the "Zeitgeist" to pass muster unsubstantiated as a causal construct. And, while they are at it, I would also appreciate explanations of the astrological (and astral-traveling) beliefs of the 1993 Nobel Prize winner in chemistry, Kary Mullis (1998) and the Eastern mysticism of the theoretical physicist David Bohm (Pleat 1997).[5] It would be unsatisfactory, and unfair, to hold all natural scientists responsible for the beliefs of these colleagues, and, for the same reasons, I would hope that collective accountability would not be enforced on the social sciences and humanities. (We have our Kary Mullis's too—but then you knew that!)

Meanwhile, opportunities for constructive conversations between sociologists and concerned scientists are being missed. I say that I find ignorance of science appalling. I do not say so as a sociologist of science: dismaying (or confusing) as it may be to my scientist-friends and colleagues, sociologists doing their day-jobs (in my part of the grove) are *obliged* to adopt a neutral position with respect to the validity of the beliefs we study. That's just to say, we take it as a valuable and innocuous *maxim of method* that we describe and explain the credibility of beliefs

4. See, for example, the current General Social Survey: http://www.icpsr.umich.edu/GSS99/subject/science.htm.

5. There are some natural scientists who have professed skepticism about scientists' forays into the philosophy of science, and others who doubt whether, for example, a chemist's views of invertebrate zoology or a seismologist's view of plant biochemistry are much more informed than those of an educated layperson. Einstein's view "that the man of science is a poor philosopher" is quoted in my first-round contribution to this volume [8]; see also the confession by the awesomely polymathic Robert Oppenheimer (1955, 125) "that only by good luck and some hard work do I have even a rudimentary notion of what cooks in other parts of the house called science than the one that I live in."

as we find them, irrespective of whether—when we clock off of our day jobs—we like them or not or we find them true or false. When I say that I find ignorance of science appalling, I do so as an average educated person.[6] I do not like it any more that some of my students and neighbors are creationists than I like it if they think that England is an American state or that the United States fought Russia in the Second World War. In my view, these are all marks of personal ignorance and, indirectly, of the systematic failure of our educational institutions. And, if one may dare say so in the present climate, public ignorance of science might also have something to do with the reluctance of most scientists to commit much of their valuable time to addressing the public effectively, or with the risks younger scientists take with their careers should they actually do so.

While the day-job, or professional, neutrality of sociologists may baffle some scientists, it is, in our view, a necessary condition for trying to find out just *why it is* that some folk believe in astronomy and others astrology, some in Darwin and others in Scripture. That's what we do.[7] And, if we are any good at our work, then we *do* find out some things about the distribution and grounds of belief. One would like to think that scientists genuinely concerned about how scientific knowledge secures credibility, or fails to do so, might be interested in some of our findings. It would be good to talk to them about such matters. Let us find ways to do it, for we have shared interests in the state of the culture.

6. A break between on-the-job and off-the-job orientations is not, of course, particular to sociologists of science. It applies to physicists as well, as the example of Eddington's Table establishes: acting as if they know it's mostly empty space while on the job; acting as if they know it's solid and durable while off the job.

7. To be sure, many sociologists are also interested in why some *scientists* believe, for example, in Darwinian gradualism and others, equally well credentialed, in punctuated equilibrium theory. But the grounds of credibility are what most of us are concerned with, wherever variation in beliefs happens to be socially located. Many scientists are engaged in struggles for credibility *within* their disciplines, and I can imagine practical reasons why some of them might also be interested in such sociological "internal" studies.

Chapter 24

PEACE AT LAST?

Steven Weinberg

In reading the other essays in this book I was disappointed to see how much there was in them with which I could agree. The essays by sociologists, historians, and philosophers did not criticize the way that science is done, any more than the essays by scientists objected to the study of science as a social enterprise. We seem to be hurtling toward a general reconciliation. But perhaps it is still not too late to draw back from the brink of peace.

It seems to me that there is at least one important issue left to be argued. Although scientists recognize that their theories often bear the stamp of the social environment in which they are formulated, we like to think of this as an impurity, some slag left amid the metal, which we hope eventually to eliminate.[1] We have felt the powerful attraction that true theories exert on our thinking, an attraction that seems to have little to do with the social setting of our research. We wonder why some historians of science like Harry Collins are not interested in describing this process, the often slow and uncertain progress of physical theories toward an ultimate culture-free form that is the way it is because this is the way the world is.

In a 1996 article (Weinberg 1996) I tried to express this view of science by saying that when we make contact with beings from another planet we will find that they have discovered the same laws of physical science as we have. In his essay in this book, Michael Lynch [4] very properly caught me up on this and pointed out that the scientific works of the

1. Well, at least physicists. Or at least some physicists. Or maybe just me.

intelligent creatures who inhabited Earth until a few centuries ago did not resemble the theories we believe in today. But this just underlines the point that I am trying to make here, that it is not necessarily the state of scientific theory at the moment that is culture-free, but its asymptotic form, the form toward which it ultimately tends. As I understand it, most sociologists of science either deny the existence of this asymptotic limit or choose to ignore it.

In arguing about this, scientists are likely to mutter about "truth" and "reality." We talk about the truth of our theories and how they correspond to something real. This is a dangerous business; philosophers have been arguing for centuries about what is meant by truth or reality. Ian Hacking (1999) calls these "elevator words," by which he means that they are used to elevate the importance of what we are talking about, without really being of much help in settling anything. I have myself expressed the opinion that when we say that a thing is real we are simply expressing a sort of respect (Weinberg 1992) [32]. In a more negative spirit, Thomas Kuhn has said that "no sense can be made of the notion of reality as it has ordinarily functioned in the philosophy of science" (1992). So it is not surprising that both Michael Lynch [4] and Steven Shapin [8] in their essays in this book and Richard Rorty elsewhere (1997) took issue with my statement that "for me as a physicist the laws of nature are real in the same sense (whatever that is) as the rocks on the ground."

Of course in saying this I did not think that I had solved the ancient ontological problems surrounding the concept of reality. This is why I inserted the parenthetic clause "whatever that is" in speaking of the sense in which rocks are real. Maybe I should have been more explicit in my modesty, which Shapin took as nervousness.

On the other hand, I do want to defend the use of words like "truth" and "reality" in discussing the history and sociology of science. Everyone uses these words in everyday life: the monster in my dream last night was not real, the deer on the lawn this morning is real, and that's the truth. We do not have a precise conception of what we mean by these words, any more than we have a precise conception of what we mean by other words like "cause" or "love" or "beauty," but they are still useful to us. When we say that something is real we intend to convey that our experience of it has one or more of those properties that give us the impression that something has an independent existence: maybe other people can experience it; maybe it doesn't change when our mood changes; maybe it is something about which it is possible to make mistakes; maybe it doesn't go away when we look at it more closely. Individual rocks are eminently real in this sense, though the category "rocks" may not be. My comparison of the laws of nature to rocks was not a philosophical argument, but rather a personal report, that my experience of the laws of nature in my work as a physicist has the same qualities that in the case of rocks make me say that rocks are real.

This raises a larger issue. As shown by our common use of words like "real" and "true," we all adopt a working philosophy in our everyday lives that can be called naïve realism. As far as I know, no one has shown why we should abandon naïve realism when talking of the history and sociology of science. Philosophers may be able to help us to sharpen the way we understand words like "real" and "true" and "cause," but they have no business telling us not to use them [29].

Part Three

REBUTTALS

Chapter 25

REPLY TO OUR CRITICS

Jean Bricmont and Alan Sokal

Before addressing the substantive criticisms directed at our essay, it is necessary to set straight a few commentators' misreadings of our arguments.[1]

According to Steven Shapin, "Bricmont and Sokal attribute the cultural credibility of flat-Earth or witchcraft beliefs to 'the existence of a radically relativist academic Zeitgeist'" [23]. But we did no such thing. Rather, we noted "the existence of a radically relativist academic Zeitgeist" in which some "otherwise reasonable researchers or university professors . . . will claim that witches are as real as atoms"—their obvious intent being to cast doubt on the existence of atoms, not to assert a sincere belief in the existence of witches—"or *pretend to have no idea* whether the Earth is flat, blood circulates, or the Crusades really took place" [emphasis added]. We were thus discussing extreme *relativism or skepticism* in academia; we made no reference whatsoever to extreme *credulity* in the general nonacademic culture, much less did we claim that the latter is a causal consequence of the former.

Jane Gregory says that in our brief discussion of public-policy issues involving science, "Bricmont and Sokal suggest that we need to know how the world really works *in order to determine* whom we should trust" [17, emphasis added]. But she gets our point exactly backward. In setting public policy with regard to (for example) BSE, nuclear power, or global warming, it is desirable to have an as-accurate-as-possible understanding

1. We prefer to pass over without comment the pejorative epithets directed at us: "noisiest guests at the party" (Pinch); "sandlot philosophy" (Mermin).

of the underlying natural phenomena (i.e., how the world really works).[2] But because politicians and citizens frequently have neither the time nor the expertise to evaluate the scientific evidence themselves, they are obliged to do something *second-best:* decide which experts to trust. We have no magic recipe for how best to do this, but we did suggest a few guidelines, based on epistemological and sociological considerations.

Now to the substantive issues, the foremost of which is methodological relativism.[3] Harry Collins [15] begins by setting forth admirably clear definitions, in terms nearly identical to our own, of "ontological relativism" and "epistemological relativism"—any combination of which he calls "philosophical relativism"—and "methodological relativism." In particular, he defines methodological relativism as the injunction that "the sociologist or historian should act as though the beliefs about reality of any competing groups being investigated are not caused by reality itself."[4] But then he asserts, misleadingly, that our own "central argument [is] that methodological and philosophical relativism cannot be disentangled." Quite the contrary: we *disentangle* (i.e., distinguish) the two doctrines exactly as Collins does. Our central thesis is, rather, that if the sociologist's aim is to give a causal account of some individual's or group's beliefs—as, for instance, the Strong Programme aspires to do—then methodological relativism cannot be *justified* unless one also adopts philosophical relativism or radical skepticism. Since we have presented the arguments for this thesis in great detail in our essay [3] and again in our commentary on other contributors' essays [14], we need not repeat the reasoning here. But we do need to set straight Col-

2. It should, we hope, go without saying that even a complete understanding of the scientific facts would not, by itself, suffice to determine *what to do:* policy decisions will inevitably involve political, economic, and ethical considerations as well as scientific ones. (In Bayesian terms, it is necessary to specify the utility function and not merely the posterior probability distribution.)

3. Let us mention in passing our irritation at Trevor Pinch's condescending comment that "methodological relativism [is] a term which *these days* Jean Bricmont and Alan Sokal *have learned to use*" [21, emphasis added]. For what it's worth, we've been using this term ever since our earliest essays on the subject—see, for example, Sokal 1998 (written in early 1997), Bricmont 1997, and Sokal and Bricmont 1998 (first written in French in June 1997)—and we challenge Pinch to find any publication of ours in which we've used any *other* term for this concept.

4. This definition is plagued, however, by one key ambiguity: Is the investigator being told to act as though the beliefs are not caused, *even in part,* by reality itself? Or merely to act as though the beliefs are not caused *solely* by reality itself? The distinction between the two doctrines is crucial, for as we have argued in our essay, the first doctrine is untenable (unless one adopts philosophical relativism or radical skepticism), while the second is eminently sensible. Collins's note 1 [15] suggests that he may intend the first interpretation: "I set out this view [methodological relativism] most clearly in a paper published in 1981, in which I said: '[T]he tenet of symmetry implies that we must treat the natural world as though it in no way constrains what is believed to be.'" However, even this latter formulation is ambiguous, for it depends on how one interprets the verb "constrains."

lins's misinterpretation of one of our arguments against methodological relativism, namely, the *reductio ad absurdum* concerning the inverse-square law.[5] Collins portrays our thought experiment as an exercise in counterfactual reasoning: "arguments based on how things would have turned out if we could have altered some aspect of the past."[6] But we did *not* propose to alter any aspect of the past; rather, we proposed an experiment that could in principle (albeit probably not in practice) be carried out *today or at some time in the future*. The question is not what *would* have happened in the seventeenth century if no information about planetary movements *had been* available; it is whether one could conceivably explain what *did* happen (as well as what *did not* happen) without making reference to the information about planetary movements that *was* available in the seventeenth century.[7,8]

Nor did we "pick an easy case," as Collins asserts, by selecting an example in which the underlying scientific controversy was settled long ago; the reasoning would be identical were one to select a currently raging controversy such as global warming. The question is: Could one conceivably explain scientists' beliefs about the Earth's climate without making *any* reference to the currently available *evidence* concerning the Earth's climate?[9]

5. Let us note, in passing, that we gave *three* versions of the argument against methodological relativism—"it is raining today," Newtonian mechanics, and the *reductio ad absurdum* concerning the inverse-square law—and Collins ignored entirely the first two.

6. We do not wish to enter here into the delicate debate concerning the extent to which counterfactual reasoning is legitimate in historical inquiry. But we note with amusement that Peter Dear [16] advocates precisely the type of counterfactual history that Collins rejects.

7. Of course, it can be argued that any historical "explanation" necessarily relies (implicitly or explicitly) on some causal theory and that any causal theory necessarily entails numerous counterfactual assertions. To the extent that these theses are valid, our thought experiment *would* implicitly involve counterfactual reasoning; but then Collins's objection to such reasoning would be without force.

8. Let us note in passing that, pace Collins, we do *not* assert that belief in astrology can be explained in purely sociological terms, nor that it could be explained "without reference to the movement of the planets." We say only that "[i]n this case it is *at least conceivable* that one could obtain a purely sociological or psychological account of the incidence of such beliefs, without ever invoking the *good evidence supporting those beliefs*—simply because there is no such evidence" [emphasis added]. Obviously any valid explanation of belief in astrology would have to make reference to the movement of the planets, if only because those movements play a central role in astrological beliefs; all we assert is that the explanation would make no reference to the alleged *causal processes* by which the planets influence human life (according to astrological doctrine), for the reason that those causal processes are, in our best rational judgment, nonexistent. Note also our observation that "if you happen to believe (wrongly) that astrology *is* well supported by evidence, then this factor *should* presumably enter into what you regard as a satisfactory causal account of belief in astrology."

9. Collins is perhaps confusing this issue with a *different* one in which the distinction between long-settled science and unsettled controversies *is* relevant, namely: how the sociologist or historian of science should act when studying controversies on which he

David Mermin also misunderstands us on this issue [20]: he asserts that we "often . . . explain the emergence of truth by reference to its truthfulness." But that mischaracterizes our reasoning, which is to make the elementary observations that (a) in at least some cases, people believe statements in part because of *evidence* that those statements are at least approximately true; and (b) the existence of such evidence is often causally linked to the fact that the statement *is* at least approximately true. Moreover, in many cases, *after* one has obtained strong evidence for the (approximate) truth of the underlying proposition, one is also enabled to understand, at least in part, the causal processes at work in (b).[10]

---

or she does not have the scientific competence to make an independent assessment of whether the experimental/observational data do in fact warrant the conclusions the scientific community drew from them. (This situation is particularly likely to arise when the sociologist is studying contemporary science, since in this case there is no other scientific community besides the one under study that could be relied on to provide such an independent assessment. By contrast, for studies of the distant past, one can take advantage of what subsequent scientists learned.) In such a situation, the sociologist will be understandably reluctant to say that "the scientific community under study came to conclusion X because X is the way the world really is"—*even if it is in fact the case* that X is the way the world is and that is the reason the scientists came to believe it—because the sociologist has no independent *grounds to believe* that X is the way the world really is other than the fact that the scientific community under study came to believe it. Of course, the sensible conclusion to draw from this cul de sac, it seems to us, is that sociologists of scientific knowledge should abstain from studying scientific controversies on which they (together with their scientist collaborators, if any) lack the competence to make an independent assessment of the scientific evidence, if there is no other (for example, historically later) scientific community on which they could justifiably rely for such an independent assessment.

(To forestall any possible misunderstandings, let us stress that we are *not* telling sociologists that they must refrain from studying contemporary scientific controversies. We are only saying that *if* they aspire to study the substantive content of scientific controversies [and not merely the social structures of the scientific community], and *if* they want these studies to be logically sound, *then* they cannot avoid making an independent assessment of whether the experimental/observational data do in fact warrant the conclusions that various scientists drew from them; moreover, the sociological conclusions of their study will be only as reliable as this independent assessment of the scientific evidence. For it is not enough to study the alliances or power relationships between scientists, or their deployment of rhetoric, important though these aspects may be: what appears to a sociologist as a pure power game may in fact be motivated by perfectly rational considerations which, however, can be understood as such only through a detailed understanding of the scientific theories and experiments. Consequently, sociologists or historians who aspire to analyze the substantive content of scientific controversies need to possess—or acquire—the scientific competence to make an independent assessment of the evidence, or else work together with scientist collaborators on whom they can rely for such an assessment. Of course, many historians of science, and some sociologists of science, *do* possess this competence in the scientific subfields they study; and many others are capable of acquiring enough scientific competence to collaborate fruitfully with scientists in interdisciplinary teams. We thank Harry Collins and Jay Labinger for fruitful misunderstandings that spurred us to write this clarification [35].)

10. Thus, our belief that the Earth is approximately spherical is due in part to the fact that it *is* approximately spherical: for if it were (for example) flat or tetrahedral, present-day observation techniques would allow us to know that.

Collins summarizes our view, albeit a bit crudely, by saying that beliefs[11] can be divided roughly into three classes: "type 1, where the current consensus is overwhelmingly dominated by the natural world . . . ; type 2, where the outcome is overwhelmingly dominated by the social world . . . ; and type 3, perhaps live controversies, where there is a mixture of both causes."[12] And he correctly states our view that "even in the case of type 3 science, methodological relativism is untenable [unless one also adopts philosophical relativism or radical skepticism] because if you know there is some contribution from the natural world you must include it." But then he purports to refute our view by *reductio ad absurdum,* as follows: "Thus, in type 3 cases there are two types of cause. Therefore in type 3 passages of science, explanatory papers must invoke both kinds of cause if they are sincere and free from error. This means that papers in type 3 areas of science, written by scientists and published in a scientific journal, that did not mention the social factors that contributed to the scientists' beliefs would be flawed." But this manifestly confuses papers whose goal is to present arguments for or against the underlying scientific propositions with papers whose goal is to explain scientists' beliefs. Most papers published in scientific journals are of the former type, and it would be silly for them to drag in social factors that can have no causal effect on the scientific phenomena under discussion.[13]

Collins offers a second purported *reductio ad absurdum* of our view: he notes that the inverse-square law would not likely have become accepted "unless humans had brains of a certain size and unless civilization had not been destroyed by meteor impact prior to the law's discovery. Yet the study of the origins of our belief in the inverse-square law does not have to involve either human anatomy or terrestrial catastrophes." But Collins forgets that our thought experiment asked one to explain why

11. We use the neutral word "belief" instead of the word "science" employed by Collins, because some of the beliefs in question (e.g., astrology) cannot, in our view, be properly characterized as being "science" (to put it mildly). Let us hasten to add that we do *not* assert the existence of a sharp demarcation between "science" and "nonscience," much less one based on rigid criteria such as those proposed by Popper or the logical positivists; rather, we make a case-by-case evaluation of the extent to which the beliefs in question are based on a rational evaluation of empirical evidence, and we note that astrology is radically different in this respect from the mainstream sciences.

12. Collins's summary of our view is too crude because (a) he draws too sharp a distinction between types 1 and 3, (b) he oversimplifies the complex *interaction* between natural and social factors in type 3, which cannot be reduced to a mere "mixture of causes," and (c) in types 2 and 3 it is necessary to include also psychological and biological factors.

13. On the other hand, scientific journals do occasionally publish papers of the latter type (e.g., some historical review articles), and such papers *should* bring in social factors to the extent that they are causally relevant.

seventeenth-century English physicists came to believe that the gravitational force decays with the inverse square of the distance, *rather than the inverse cube.* The nondestruction of the human race and a certain minimal size of the human brain are necessary preconditions for humans' belief in *either* the inverse-square *or* the inverse-cube law; they have no *differential* explanatory value, and thus need not be included in the requested explanation.[14]

Finally, Collins misstates Hume's advice on miracles as "do not believe anything 'miraculous' unless you see it with your own eyes." The correct formulation of Hume's (and our) view is rather: do not believe anything "miraculous" unless you see it with your own eyes *or* the proponents offer you reasons to believe the "miraculous" claim that are more compelling than the alternative explanations of fraud or self-deception.[15] Thus, as Mermin notes in his essay in this volume [7], "I have never been to China. I can nevertheless assure people that such a place exists, because the conspiracy necessary to fool me into believing it, were it false, is too implausible to contemplate." Similarly, though neither of us has personally witnessed the experimental measurement of the magnetic moment of the electron—or, for that matter, personally carried out the order-α^3 theoretical calculation—we are convinced by the truly "miraculous" agreement between theory and experiment[16]

Theory	1.001 159 652 201 ± 0.000 000 000 030
Experiment	1.001 159 652 188 ± 0.000 000 000 004

that quantum electrodynamics must be saying *something* at least approximately true about the world;[17] our judgment that this explanation is more likely than fraud or self-deception results from a web of reasons that are both scientific (the experiments we *have* witnessed, the calculations we *have* performed) and sociological (our understanding of the scientific community's epistemology, methodology, and social structure).[18]

14. The structure of the human brain *would* be relevant if it were being argued that humans have an *innate* predisposition (e.g., one sculpted by natural selection) to believe in the inverse-square law rather than the inverse-cube. This is, of course, unlikely in the case at hand, but it could well be relevant to some other scientific theories, for example, to humans' predisposition to believe in Euclidean rather than Lobachevskian geometry.

15. Of course, one should also be duly skeptical of "miracles" observed with one's own eyes, and take account of the possibility of faulty perception, misinterpretation, etc.

16. See Kinoshita 1995 for the theory, and Van Dyck, Schwinberg, Dehmelt 1987 for the experiment. Crane 1968 provides a nontechnical introduction to this problem.

17. In a sense of the phrase "approximately true" that does *not* require the approximately true theory to have the correct *fundamental* ontology.

18. Of course, to spell out in detail the reasoning underlying this judgment would be rather lengthy. See, for example, Haack 1998, chapters 5 and 6, for a brief sketch of some of the factors that would enter.

Peter Saulson [22] is disappointed that our essay concentrated (as we made clear at the beginning that it would) on SSK's declared methodology rather than on its concrete case studies.[19] All we can say is, *à chacun son goût:* both are legitimate objects of analysis.[20] In particular, since the "new" sociology of science distinguished itself from its Mertonian predecessors precisely by trumpeting its own radical methodology (embodied notably in the symmetry principle), it is of especial interest to analyze critically that proclaimed methodology.

Saulson does make one valid criticism: he notes that our analysis of the Strong Programme's methodological precepts was incomplete because we omitted addressing Bloor's statement in principle 1 that "Naturally there will be other types of causes apart from social ones which will cooperate in bringing about belief." This question was addressed in our book, where we observed that "the trouble is that he [Bloor] fails to make explicit *in what way* natural causes will be allowed to enter into the explanation of belief, or what precisely will be left of the symmetry principle if natural causes are taken seriously."[21] It is ironic to note that Bloor's position on the causal role of the natural world has been criticized, in terms far harsher than our own, by Bruno Latour, who sees it as mere lip-service aimed at fending off the accusation of idealism: "[A]re these objects allowed to *make a difference* in our thinking about them? The answer given by David [Bloor] and repeated over and over again by all the descendants of this tradition—even empirically minded ones such

19. Collins [15] also finds our choice disappointing, especially inasmuch as we acknowledged that "interesting work may have been done" in some of the concrete studies carried out by SSK practitioners. And Collins goes on to compare us to people who would claim that "although quantum theory has come up with useful results, the underlying indeterminacy is impossible to accept, and . . . quantum theorists should go back to doing sensible classical physics." But this comparison is ludicrous. First of all, when we said that "interesting work may have been done" in some of the empirical studies associated with the Strong Programme, we never meant to concede (and certainly do not believe) that those studies enjoy the same level of empirical support as does quantum mechanics. (Did we perhaps overlook some recent successes in SSK comparable to the eleven-decimal-place agreement between theory and experiment achieved by quantum electrodynamics?) Second, for the reasons explained in our essay, nothing in those empirical studies justifies methodological relativism (much less philosophical relativism, as Collins used to think: how could any empirical study justify a philosophy that basically says that empirical studies never allow us to discover objective truths?). Finally—though this is a much subtler point, which we cannot address in detail here—even in the case of quantum mechanics, it is far from obvious that its empirical successes warrant the philosophical claims usually associated with the Copenhagen interpretation.

20. Perhaps the difference in taste arises in part from the fact that we are theoretical physicists, while Saulson is an experimentalist. See Gingras 1995 for a critical analysis of several recent tendencies in SSK, focusing on concrete case studies; see also some of the essays in Koertge 1998.

21. Sokal and Bricmont (1998, footnote 115, emphasis in the original). See Laudan (1981) and Slezak (1994a) for more detailed analyses of Bloor's ambiguities.

as Shapin, Schaffer and Collins—is a resounding 'no'" (1999a, 117, emphasis in the original).[22] Finally, although Bloor is vague about the role that the natural world should play in sociologists' explanations of scientists' beliefs, other SSK practitioners (such as Harry Collins) are explicit in their rejection of such a role, as discussed above.

Concerning redefinitions of truth in terms of intersubjective agreement, we argued in our essay [3] that "once upon a time, people agreed that the Earth was flat (or that blood was static, etc.), and we now know that they were wrong. So intersubjective agreement does not coincide with truth (understood intuitively)." Mermin [20] criticizes this argument by saying, "The fact that 'once upon a time, people agreed that the Earth was flat . . . , and we now know that they were wrong' does not establish that there is more to objectivity than intersubjectivity. All it shows is that in this instance the inference of the objective from the intersubjective was inadequately based or badly drawn." But Mermin's formulation implicitly concedes the point that we were trying to make: for if it is possible to make an inadequately based inference from A to B, then clearly B cannot be *identical* to A.[23]

Mermin [20] says that we "exaggerate" when we attribute to Barnes, Bloor, and Henry (BBH) "a rather tolerant (or even favorable) attitude" toward astrology, but he gives no evidence to support this charge.[24] He

22. Let us stress that we do not agree with most of Latour's criticisms of the Edinburgh school.

Let us also note in passing the double standard upheld by Peter Dear [16], who defends Latour's criticisms on the grounds that "such observations take us inside science studies, rather than conforming to caricatures of science studies drawn by unengaged critics"—and this even though our own "caricature" of Bloor's views is rather more charitable than Latour's "inside" critique.

23. Of course, it could turn out that B is *logically equivalent* to A, by virtue of a deductive argument of which the people in question are unaware. But that possibility is not relevant to the case at hand, in which A (intersubjective agreement that the Earth is flat) is true while B (flatness of the Earth) is false.

Note also that Mermin's use of the word "objective" confuses two distinct questions: whether the belief in question is (objectively) *true,* and whether it is (objectively) *rationally justified* with respect to the available evidence. We contend that *neither* truth nor rational justification is equivalent to intersubjective agreement, but different arguments are needed to establish the two claims. (In particular, the flat-Earth example does *not* suffice to prove that rational justification is different from intersubjective agreement, since the belief in a flat Earth may well have been rationally justified in all those places and times where it was widely accepted.)

24. Indeed, Mermin himself cites, in his review of the BBH book (Mermin 1998a, 621–22), the relevant passage: "Astrology . . . and homeopathy . . . remain firmly saddled with the label of pseudo-sciences in spite of recent work which seems to some to call for a reassessment (Gauquelin, 1984; Benveniste, 1988). Michel Gauquelin's statistical evidence in support of astrology would perhaps be a serious embarrassment to scientists if they were not so good at ignoring it. But one day it could conceivably come to be accommodated as a triumph of the scientific method. Gauquelin's work seems to imply

further protests that his Humean argument for the implausibility of Gauquelin's alleged data was not intended "to refute an argument for astrology, but [only] to expand on Bloor's remark that [BBH's] point was well illustrated by my own unwillingness to take off a year or two attempting to duplicate the data claimed by Gauquelin." We accept Mermin's statement of his intentions, but his reasoning nevertheless shows that astrology is so unlikely to be true that (a) it's not worth believing in astrology unless its advocates come up with vastly more convincing empirical evidence, and (b) it's not worth taking off a year or two to study the evidence unless astrology's advocates come up with vastly more convincing empirical evidence. Perhaps Mermin intended only to assert (b), but the same "Humean" reasoning leads to both (a) and (b).

An even starker example of a sociologist's favorable attitude toward a pseudoscience—in this case homeopathy—is provided by Jane Gregory's characterization of Benveniste's claims about the memory of water as "inconvenient truths" [17]. In order to illustrate why we (and most scientists) take a vastly more skeptical attitude toward homeopathy and astrology than do some sociologists, we would like to discuss critically some of Gregory's assertions.

Gregory's central claim is that "[i]n science, replications, peer review, and publication in *Nature* are usually good enough: the end product is usually well on its way to becoming what Bricmont and Sokal might call 'reality' or 'truth.'" To begin with, this grossly misunderstands what we mean by "truth": as we explained at length in our essay, "truth" signifies for us "correspondence with reality"; it thus makes no sense to say that an assertion *becomes* true through replication, peer review, and publication. But more importantly, while "replications, peer review, and publication in *Nature*" can constitute *evidence* (sometimes strong evidence) for the truth of a scientific claim, they are by no means conclusive, nor are they the sole criteria for judging a claim. And when the claim concerns the alleged "memory of water," even a few replications, peer reviewed and duly published in *Nature,* do not suffice to overcome our rational skepticism. Why not? The crux of the matter is not, as Gregory seems to think, that Benveniste's claim is an "inconvenient truth"

the existence of forces and interactions unrecognized by current scientific theory and yet it is based on methodological principles and empirical evidence which have so far stood up to sceptical challenge" (Barnes, Bloor, and Henry 1996, 141). Though this passage does not indicate unequivocal support for astrology, it does demonstrate a tolerant (and even cautiously favorable) attitude toward astrology, as well as a failure to comprehend the vast gulf between the established natural sciences and astrology as regards both methodology and degree of empirical confirmation. See also note 28 below.

that threatens to "mak[e] redundant the pharmaceutical industry"; it is rather that everything we know about physics and chemistry renders this claim so improbable that an overwhelming quantity and quality of evidence would be needed in order to make it believable. This attitude—namely, "the more implausible the claim, the stronger the evidence needed to justify it"—is of course common sense, but it is not for that reason any less valid.[25] Indeed, all human beings necessarily proceed in this way: while one or two witnesses might suffice to convince Gregory that Baroness Thatcher dined last night with General Pinochet, even a hundred witnesses claiming to have seen Thatcher dine in a flying saucer would be taken with a grain of salt. Likewise—though it would be too long to explain the reasoning in detail here—the idea that substances may have therapeutic effects (other than the placebo effect) even when they are "diluted" so much that not a single molecule of the original substance remains in the final product *does* run counter to all modern physics and chemistry (based, as they are, on the atomic theory of matter).

One may then ask which is more probable: that a peer-reviewed paper published in *Nature* is wrong, or that the whole edifice of modern physics and chemistry is badly flawed? And the sensible response is again provided by Hume's argument against belief in miracles. We know, from direct experience, that articles published in respectable scientific journals (such as *Nature*) can be wrong. But we have no evidence at all in favor of the claims made by homeopathy,[26] and rather heavy evidence against it (namely, all the experimental evidence confirming modern physics and chemistry).

Of course, science is not a religion:[27] its claims, even the best-established ones, are in principle revisable, so that the "memory of water" could turn out to be a real effect after all (and due to hitherto unknown causes). But this remark applies to all pseudoscientific claims, even astrology.[28] Moreover, all the pseudosciences are sometimes sup-

25. For what it's worth, this precept can also be justified by an elementary Bayesian calculation.

26. Not even in favor of what Gregory calls "the potency of homeopathic remedies," since, to our knowledge, that alleged "potency" has never been established by means that could convince a reasonable skeptic, e.g., through double-blind experiments.

27. We therefore cannot share Peter Saulson's avowed belief in "the Church of Science" [22]. For further discussion of the radical methodological opposition between science and religion, see Bricmont 1999.

28. However, as Mermin observed, "BBH's gloss on astrology—'the existence of forces and interactions unrecognized by current scientific theory'—fails adequately to convey the truly spectacular degree to which compelling evidence in support of astrology would require a massive radical reconstruction of our current understanding of the world" (Mer-

ported by alleged spectacular discoveries in their favor (Benveniste, Gauquelin, etc.). But if we adopt a nondiscriminatory attitude with respect to all similar claims—faith healing, New Age medicine, and the lot—how are we supposed to proceed? There is, after all, a vast number of such theories (especially if, following our nondiscriminatory policy, we include the traditional beliefs of other cultures). How much time and effort should one invest checking each of those claims? The most reasonable attitude, it seems to us, is to do what most scientists do: namely, keep a skeptical eye on those "miraculous" claims, wait and see if some really convincing evidence turns up some day, and then, if it does, find out what has to be revised in the standard scientific worldview.

To say, as Collins does in the context of our discussion of Benveniste's claims, that for us "replication of observations is the key" is not quite right. In the case of highly implausible claims such as the memory of water, even a few replications are insufficient; Hume's argument still applies. Mermin said it well: "An important motive behind rejecting such claims without any attempt at replication, unmentioned by BBH but clearly recognized by those doing the rejecting, is the gross inefficiency of investing extensive time and resources in an attempt to refute overwhelmingly improbable claims. For similar reasons, one turns down an offer, rendered on the spot, to purchase the Brooklyn Bridge for five dollars, without making a trip to the courthouse to confirm the conjectured non-existence of the claimed deed of ownership" (1998b, 642).[29]

Finally, though in this reply we have necessarily focused on disagreements (and on misrepresentations of our views), we wish to conclude by noting briefly some important points of agreement with other contributors:

- We share Shapin's equal distaste for ignorance (or impermeability to empirical evidence) in biology and in history.
- We applaud Lynch's disapproval of ad hominem argument.
- We second Pinch's observation on the diversity of views in this debate, which are by no means determined by one's profession.
- We agree with Saulson that "[o]utside observers with a background in understanding social life clearly have a primary role to play in elu-

min 1998b, 642). A similar remark can be made for homeopathic claims (although the reconstruction might be less radical in this case).

29. We do, however, wish to apologize to Collins and Pinch for misunderstanding their assertion that "if homeopathy cannot be demonstrated experimentally, it is up to scientists, who know the risks of frontier research, to show why" (1993, 144). We accept their assurances [15] that they were not defending homeopathy or attempting to shift the burden of proof away from homeopathy's advocates.

cidating the function of science as a social system, embedded in a larger society."[30]

- Last but not least, we concur with Mermin that "no reliable body of knowledge can be undermined by viewing its acquisition as a collective human activity," and with Dear that "a detailed investigation of the sociohistorical conditions that brought about belief in some scientific truth-claim does not in itself undermine that claim."

30. However, we disagree with Saulson's next paragraph, where he asserts [22] that "many scientists are prepared to dispute" this view, without citing a single example. For what it's worth, we have repeatedly stressed our belief in the value of an intellectually rigorous sociology of science. Indeed, even the reputed "extremists" Gross and Levitt stated in no uncertain terms that

> Natural scientists . . . do not feel that their particular expertise in some area of science automatically endows them with insight into the human phenomenology of scientific practice, or that familiarity with the recent results and the liveliest questions of their specialty qualifies them to pronounce on its evolution as that relates to the course of human development. Apart from the most arrogant, they concede that the psychological quirks and modes of personal interaction characteristic of working scientists are not entitled to special immunity from the scrutiny of social science. If bricklayers or insurance salesmen are to be the objects of vocational studies by academics, there is no reason why mathematicians or molecular biologists shouldn't sit still for the same treatment. (1994, 42)

Chapter 26

CROWN JEWELS AND ROUGH DIAMONDS: THE SOURCE OF SCIENCE'S AUTHORITY

Harry Collins

IS THERE ANYTHING TO WORRY ABOUT?

It seems to me that I have already anticipated the points made by Bricmont and Sokal in their second-round piece [14] in my second round [15], so I need not return to the topic of "complete explanations." I am curious to know how Steve Weinberg will respond to our disagreement about history, but whatever he says I will not have an opportunity to reply, so I will leave that topic too.

But I will take my cue for this final round from Weinberg: he is worried that we are all agreeing too much [24]. I certainly would not want it to be thought that there is "nothing to worry about" in the sense that the social sciences are saying so little that there is nothing worth thinking through. What I will do here, therefore, is to try to set out what is new and "dangerous" about the project of SSK. That is, I will set out the things that SSK does that I think are good for science but which might seem less than attractive to some scientists.

The big disagreement, if there is one, ought to be about the source of science's authority. I suspect this is where Weinberg and I might have disagreed most strongly at the outset of this debate, but I don't know if we still disagree.

I imagine that at the outset of this debate Weinberg would find the preeminent justification for science in its privileged access to reality, its potential for producing a "final" theory of the world, and its extensive store of knowledge about the way the world works. I would want to justify science by reference to the assiduousness, experience, skill, and virtuosity of its practitioners. In other words, Weinberg would want to justify science on the basis of its past and future achievements, I would want to justify it on the basis of the special skills of its practitioners. We can see these are not the same because even skilled practitioners often fail to achieve what they set out to achieve. This difference between us would, I suspect, be reflected in our choices of exemplary passages of science: Weinberg would choose the most polished and finished examples of the craft—he would hold up the shining product for inspection and say something along the lines of "Look what we can do if you provide the ideal circumstances and ideal questions for us." I would look at cases right at the beginning of the scientific process, where everything is murky and unsure, and I would say, "Scientists still have much to offer in spite of being forced to work with ill-formed questions under difficult circumstances where they are unlikely to achieve as much as we would like." Let us call these the "crown jewels perspective" and the "rough diamond perspective."

These different ways of justifying science are not mutually exclusive. Indeed, it may be the existence of crown jewels that makes one believe that rough diamonds are worth having. But I think that to value murky and unfinished science only because of its potential to turn into something else is risky. There are many kinds of science that cannot be polished up to a fine sparkle. Think of any science that involves a long-term historical dimension, especially one involving human action such as the science of long-term climate change. Second, we know for sure that in many cases we must make the political and policy decisions that turn on science long before even a reasonable degree of consensus has been reached among scientists. Therefore, to justify the use of science in political decision-making by reference only to polished jewels is to invite exclusion from political decision-making. We have to find out how to use murky and ill-formed science in the political process; we have to learn to value it. Dreaming of a final theory takes us away from this project. Let me hasten to add that I have nothing against crown-jewels science—it is our most wonderful cultural achievement—but it can justify itself on those grounds alone.[1]

THEORY AND EXPERIMENT

I have already made some of these points in more direct terms in my second-round piece. Most of the scientific questions faced by citizens are

1. Its budgets, remember, though large compared with other kinds of science, are still tiny compared with military spending.

of the murky and ill-formed kind, which is why they enter the political domain. My tactic has been to look at what happens when well-formed questions with potentially polished answers—such as questions about the existence and signature of gravitational radiation—are asked at an early stage. I prefer to look at well-formed questions because here science has the potential to exhibit itself at its best, so the conclusions one reaches about the scientific process cannot be ruled out by saying the questions are the wrong sort. In studying the hard sciences at their moment of formation we have removed at least one sort of confounding variable. In my view the early stages of a new passage of physics makes an excellent laboratory for studying the kind of science that is of concern to the citizen because what we learn from it will apply even if that science turns out to be perfectible.

If we look at science at these early stages we find a number of sources of potential concern for scientists who would want to preserve the vision of science that was predominant immediately after the Second World War and is still held dear by some. Experiments and passages of theorization look much less decisive than they do under the crown jewels perspective. And this is not just a matter of "looks"—experiments and theories do turn out to be less decisive in bringing scientific controversies to a close than uninvolved scientists and others generally think they are. The same applies to features of scientific method such as replication, calibration, and the use of control groups.

We are saying nothing deeply philosophical here, nor are we saying anything about the long term; we are just reporting observations—the kind of observations that scientists deeply involved in controversies will recognize. The claim is only that scientific procedures do not speak for themselves but have to be judged and interpreted. The interpretations—whom to trust and what to believe—are mediated by social relationships. This is almost too obvious to be worth stating yet it remains a conclusion that is in marked contrast to the descriptions of theory and experiment that many philosophers of science once produced. In those works theories engaged with each other in a kind of logical dance, the edges of the dance floor being marked out by robot-like machines called "experiments." This model, unfortunately, still colors the pronouncements of many of science's spokespersons.

HOW IS SCIENCE KNOWN?

Another tendency among sociologists of scientific knowledge which has the potential to irritate is the way they analyze public understanding of science. For example, not so long ago a colleague (very far from being a crown-jewels scientist himself) told me that in spite of his sympathy for sociology of science, anyone who still believed that cold fusion was a possibility must be mad. I asked him if this applied to members of non-Western cultures such as Azande tribesmen, and he agreed that it did not. That is, a member of the Azande tribe who believed that cold fusion was a possibility would not be counted as mad—merely wrong—the reason being that the Azande do not know enough to make a sound judgment. Now, suppose I say to my friend that I still think cold fusion is a possibility, eliciting the judgment that I am mad.[2] What is it that I, a relatively scientifically literate member of Western society, know that the Azande do not, that renders the judgment of insanity fair? Note that I have never done an experiment involving cold fusion, my knowledge of the theory of adsorption (I had to check the word in the dictionary) of gases in crystal lattices is slim, and I know almost nothing of coulomb forces nor fusion processes. I do, however, know someone—Martin Fleischmann—who knows a lot more about all these things than I, and he believes that cold fusion happens, so he is "mad" too. It does not look, then, as though this kind of "madness" grows out of a shortage of scientific knowledge because Fleischmann has plenty. Indeed, it would seem to me that, contrary to my friend's judgment, true madness would lie in thinking that I *did* know enough to make a scientific decision on these disputed issues: unfortunately, that criterion renders rather a high proportion of the scientific population mad too. I leave the conclusion to the reader.

What I do know much more about than the Azande know, however, is how to put on a good show in respect of cold fusion and the like among what the *Guardian* newspaper calls the "chattering classes"—in this case, the scientific chattering classes. The madness that my colleague was referring to, as with so much madness, lies in my not acting as a proper member of my society; it lies in my "stepping out of line." Or to be more charitable to my colleague's position, and to put this in a more positive way, the madness lies in my making poor *social* judgments about whom I ought to agree with rather than poor *scientific* judgments. I have made the wrong judgment about when the mainstream community of scientists has reached a level of social consensus that cannot be gainsaid in spite of the determined opposition of a group of well-qualified scientists who know far more than I do about the science.

Poor social judgments are also the problem with those who believe

2. Actually I have no opinion on the matter.

in, say, newspaper astrology *as a scientific theory*. They too are making a social mistake—they do not know the locations within our society in which trustworthy expertise in respect of the movements of the stars and planets and their influence on human life is to be found; they do not know that the correct social location is the community of astronomers not astrologers. It is no good asking either the believers in astrology or people like me who, for argument's sake, still believe in cold fusion to pay more attention to the scientific arguments, because we will find persuasive arguments on both sides, and we do not have the expertise to judge between them. We have to ask such people to believe only a subset of the interpretations they encounter in only a subset of the sources that are available to them, and we have to ask them to choose their interpretations and their sources by reference to the social locations of their origins. To cut what could be a still longer story short, the sociologist has the irritating habit of pointing out that, among the general public—and that includes scientists working outside their narrow specialization—beliefs in the validity of the findings of conventional science are grounded in the same way as beliefs in the efficacy of witchcraft among the Azande—namely, assessments of who in your society is worth believing. It's a shame, but that is how it is.

Whoa! Isn't that an argument for equal time for creationism in the schools and everything else that follows? No! But this is where the hard argument starts. In our society we believe in a certain kind of expertise—an expertise that comes out of assiduous study of a problem. We do not believe in the contents of books because they are books; we believe that the contents of books are to be believed only so long as the authors can show that they have written the books after studying the problem. We disbelieve the Jehovah's Witnesses who come to our doorstep because their arguments always refer back to a book—the Bible—rather than to the kind of assiduous study that is done in the field or in laboratories. We do not believe the creationists because their belief is disproportionately based on the contents of a book as compared with field and laboratory studies. In our society, other things being equal, we believe in the collective opinion of the group whose members have engaged in the right kind of study based on the right kind of observations. So, while my view that creationism and evolution should not be given equal time in our schools cannot any longer be based on the "heroic" picture of science that the philosophers once tried to justify, it is still based on the more modest proposition that the people who believe that evolution has substantially more going for it than creationism are the kind of people whom it makes sense to trust because they have the right kinds of skills and warrants. And this is where we come back to my disagreement with Weinberg. I prefer rough diamonds to crown jewels. Unfortunately, every now and again one comes across evolutionists whose eyes burn so brightly with scientific zeal that one does not feel they can be trusted; they believe in the book of Darwin rather than the value of scientific skills, and they do science no favors.[3]

CONCLUSION

I have offered two reasons for a certain class of scientists to worry about SSK and the like: it reduces the quasilogical authority of science and it turns public understanding of science into a matter of social education rather than scientific education. I think the worries are misplaced, and I think the change is inevitable and desirable. A science justified by its ability to produce clear and certain knowledge will be continually found wanting in the most visible aspect of its activities. A science justified by the virtuosity, experience, and assiduousness of its practitioners is far more secure. This kind of justification, let it be clear, weakens the boundary between science and other kinds of expertise. The hard work now is to put these vague notions—virtuosity, experience, assiduousness—on a more secure footing. We also need to understand how, under this model, the expertise of scientists interacts with that of other experts who may not have had complete scientific training but might have the right kinds of experience to make a contribution to technological decision-making in ill-formed areas. And we need to understand what is the right kind of expertise to count as science—for example we need to separate predominantly book-based claims, like that of the creationists, from experimentally or observationally based claims. In other words, far from there being nothing to worry about, we need to find new kinds of demarcation criteria that can sort industrial diamonds—jewels without sparkle and glitter—from the dross that surrounds them. That is much harder than separating out crown jewels.

3. The problem with cold fusion, of course, is that Martin Fleischmann and his colleagues have done experiments, not just read books. Hence we have to reach for the consensus among scientists rather than reaching our judgements by demarcating scientists as a class.

Chapter 27

ANOTHER VISIT TO EPISTEMOGRAPHY

Peter Dear

An observation of David Mermin's in round two [20] provides me with the occasion to clarify a little more the idea of "epistemography" as laid out in my original essay. Mermin writes: "Peter Dear's remark that 'modern scientific beliefs are in fact irrelevant to an epistemographical account' can be too limiting when those beliefs can give clues about the objective circumstances of historic events." There is, among other things, an important point about theory and practice here. As regards the practice of the historian of science, there is no doubt that knowledge of present-day scientific ideas often comes in very handy in wrestling with the (sometimes very alien) beliefs of the past—the more so the more recent the historical episode. That usefulness is, however, fundamentally *heuristic*. In orienting oneself with respect to a historical episode, it can happen that a more modern idea can cast an otherwise perhaps unexpected light on the material under investigation. Generally speaking, though, it is probably true to say that when this occurs, what is then seen is something that a full and scrupulous examination of the historical case should by itself serve to reveal to a decent historian of science; it is just that applying modern knowledge sometimes serves as a valuable shortcut.

But historians of science have for many years now been trained to be very wary of such shortcuts. Thomas Kuhn cautioned, in *Structure of Scientific Revolutions* and elsewhere, that the historian should, ideally, forget what he or she knows about modern science, so as to approach the historical material without potentially distorting preconceptions. Indeed, this hyperbolic recommendation was caricatured by a hostile writer in *Nature* a decade or so ago as the praise of ignorance (Harrison 1987). But Kuhn, the erstwhile physicist, was, of course, praising no such thing. He was trying to make the point that it is dangerously easy for the historian to see something apparently familiar in the scientific writings of a historical figure and then promptly to give up on looking closely enough to see that the familiarity is illusory, that it hides an interestingly *different* idea. My own remark about the "irrelevance" of current scientific beliefs to the framing of an epistemographical account was intended to restate very much the same point in a way that is, I think, logically (theoretically) sound—although practically misleading if it is taken to imply the heuristic irrelevance of modern knowledge. Indeed, I would stand very little chance of explaining (say) Galileo's ideas in mechanics to undergraduates without the resource of appealing to what they already know of post-Galilean classical mechanics, thereby to demonstrate the points of difference as well as the points of similarity between the two. This is, however, a far cry from treating Galileo as simply a step on the high road to modern mechanics, whose "mistakes" would soon be corrected by such as Huygens and Newton—the kind of picture against which Kuhn was reacting. Epistemography is unable, by its very nature, to look into a crystal ball.

I would also like to take up Trevor Pinch's imagined conversational gambit in our collective cocktail party [21] about "whether epistemography really does capture what science and technology studies is all about." As far as I'm concerned, it *doesn't;* there's a lot more to what people do in the field than just epistemography. But I wanted to suggest that epistemography is an inseparable, a *necessary* part of what science and technology studies is about—the field's sine qua non. That is why I discussed the issue of the moral foundations of methodologies. No methodology, as Harry Collins in his book *Changing Order* (1992, especially chapter 2) argued perhaps more clearly than anybody, can be completely ("algorithmically") determinative of investigative procedures. But a morally serious commitment to certain rules of practice (that is, what in some situations is referred to as adherence to the "spirit of the law") is necessary in any scholarly community. I wanted to suggest that "epistemography" captures the central commitments *shared* by scholars in science and technology studies.

An interesting point made by Jane Gregory in her remarks in round two [17] seems relevant here: "Where once we might have talked about the social responsibility of scientists, now we talk about the public understanding of science." If, as she implies, the science wars are partly the result of the fact that "science does not reflect too often or too deeply on itself," perhaps our attempts at promoting a "science peace" may have, as Mike Lynch puts it, therapeutic—moral—value for us all.

Chapter 28

LET'S NOT GET TOO AGREEABLE

Jay A. Labinger

Both Pinch [21] and Weinberg [24] comment on the level of agreement this "conversation" seems to be approaching, which Pinch notes is the hallmark of a "real" conversation. (I know nothing about the field of conversation analysis, but from my own experiences, I suspect that the finding that "real" conversations tend to end in agreement might have at least something to do with how a "real" conversation is defined. But never mind.) In this, my last opportunity to speak here, I want to offer some thoughts on the Benveniste case, which has been mentioned by a number of the commentators [3, 15, 17, 21] and also to respond to Bricmont and Sokal's criticism of my original contribution. I expect that each of my colleagues will find at least *something* they can disagree with herein.

Gregory notes that Benveniste's original work was replicated elsewhere, received positive peer review, and was published in *Nature,* which is "usually good enough" to start establishing the results as truth; and that anyone who understood these processes would be baffled by the subsequent developments [17]. I'm not sure how well that holds up even as a general description. Peer review and publication is at best tentative validation. The review process (which hardly ever includes replication) is good at weeding out claims that can be seen to be inconsistent, either internally or with previously established results, but it is not so good at detecting claims that are plausible but incorrect. So we have *many* results that make it into print and are subsequently found to be wrong. Usually that happens when someone needs to make use of the finding for further work and is unable to do so successfully, rather than because someone simply wanted to replicate the work. (At least, that's the case in my field, chemistry; perhaps it's significantly different in other areas of science, but I'd be surprised if so.)

Leaving generalities aside, though, I believe that someone who understands the "truth-generating" processes would also see how atypical (I hesitate to say abnormal, in the face of all the arguments over what counts as normal science) the details of the Benveniste case are and quickly overcome bafflement. All of the putative warranties for reliable truth—replication, peer review, and publication in a prestigious journal—were compromised at best. Replicability was by no means 100 percent—sometimes the experiment worked, sometimes it didn't, even in the principal investigators' own lab. The referees (according to an editor's note at the end of the original paper; their actual reports are unavailable to me) did *not* believe the results but were unable or unwilling to debar publication in the absence of unambiguous proof of error. And the journal editor's reservations were documented *twice* in the issue containing the paper, as well as in the report on the subsequent investigation he led (more on that in a moment).

Most disconcertingly, Gregory seems to feel (and I admit I may be misreading her here) that the work's great potential significance—"challenging our understanding of many areas of physics and chemistry"—is another reason to be surprised at its rejection. *Au contraire:* accepting the truth of the memory of water would not just challenge but require us to discard wholesale quantities of quite well established science—what some call "textbook science"—such as the nature of molecular motions in liquids. Call me reactionary if you will, but I, for one, am not going to even *begin* to think about doing that without much more compelling evidence than has been offered. Rather, I'll assume that something is wrong, even if I can't see exactly what it is—much as if I were to see a demonstration of a purported perpetual motion machine. When experiments sometimes give negative results and the authors can offer no explanation why that should happen, it doesn't require much of a leap of the imagination to suppose that the *positive* results are artifactual. (Recent similar episodes with which I've been somewhat familiar—polywater, cold fusion—have reinforced my belief in that strategy.)

I do agree with Collins [15] and Pinch [21] that Benveniste was treated quite inappropriately, and unfairly, by the editor of *Nature* and his debunking team. Their report (Maddox, Randi, and Stewart 1988) does identify a number of problems with the work—unreliable replicability, inattention to statistical issues, possibly slipshod experimental protocols—but all of this could, and should, have been examined by experts, not the self-admitted "oddly constituted group" that actually made the visit. Indeed, it should have been done *before Nature*'s grudging acceptance of the work for publication. The actual event, including such flamboyancies as taping a sealed envelope containing the encoding of samples to a ceiling (which of course implies, without actually charging it, suspicion of outright fraud), completely sensationalized the negative report. Benveniste felt he was deliberately set up by this sequence of publication followed by investigation (Benveniste 1988), and it's hard to disagree. Nonetheless, the "marginalization" of Benveniste has, in my opinion, much more to do with the contrast between the marginal character of his reported observations and the major significance of the claims he based on them.

Turning to Bricmont and Sokal: they dismiss my analogy, between scientists examining the effect of one variable at a time and sociologists excluding the actual facts as a possible cause of belief, as "absolutely wrong" [14]. I freely concede my analogy is not perfect—what analogy is? The one they counter with, that the sociologists are doing something like investigating causes of lung cancer without asking about smoking, is at least equally flawed. On an obvious level, the latter alludes to a statistically based study where effects of such variables can be accounted for explicitly and quantitatively: how can that be compared straightforwardly to a nonquantitative examination of a single case? But that's a trivial point.

Much more important is the point both I [18] and Collins [15] raised in our commentaries: the convictive force of arguments based on extrapolation from "easy" to "hard" cases is questionable. Yes, we all know lung cancer is related to smoking, just as we all know wine doesn't turn into blood. But sociologists of scientific knowledge (as I understand it) are interested in *how it happens* that we all come to know such things, which means they focus their studies on the times (or places, I suppose, for the wine/blood example) when we *didn't* all know them, and hence to treat them as known would potentially be distorting. This is obvious, as Collins notes, with respect to studies of ongoing controversies where the "truth" of the matter is still unknown. One *could* argue, I suppose, that such studies are premature and useless, but I don't hear anybody making that argument. Therefore, a study of a historical case done *as if* it were similarly unresolved should also be reasonable. At least, Bricmont and Sokal's charge that so doing amounts to inferior methodology or tacit commitment to philosophical relativism or both seems to me quite unreasonable. If I may indulge in another (shaky) analogy, it's as if Bricmont and Sokal's lung cancer studiers were to be criticized for failing to control for a factor that is not known to operate, say, left-handedness.

On the other hand, many scientists and traditional historians of science *will* look at historical episodes precisely from the viewpoint Bricmont and Sokal favor: the (currently accepted) truth of the matter and how it came to be accepted in terms of rational reactions to empirical evidence. They are engaged in a different project. To be sure, though, these projects overlap, and here I agree with Bricmont and Sokal that some of the claims made by SSK reflect little of the modesty that would seem to be logically entailed by the "bracketing out the facts" approach.

No doubt a lot of this has to do with rhetoric—what Collins calls the power of words. Bricmont and Sokal's suggestion that a simple name change could eliminate all their concerns seems to be giving a little *too* much power to words. Would methodological relativism carried out under the banner of the "Weak Programme" no longer require a commitment to philosophical relativism for logical consistency? Does ontology recapitulate philology?

However, this problem can go beyond rhetoric to substance. I'll close by citing a couple of examples (neither is particularly egregious, to be sure) from the cold fusion saga. In his dialogue with philosopher William McKinney, Pinch remarks (1999), as practitioners of SSK have always emphasized, on the difficulty of assessing epistemology and methodology in the midst of an active controversy and illustrates with one of the claims used against Pons and Fleischmann, that their cells were not stirred properly. He criticizes McKinney for not taking into account possible responses to this and other such attacks and states, "Pons and Fleischmann were able to use a dye tracer to *show* that bubbles of gas produced in the reaction stirred the cells *adequately*" (emphasis added). If that is intended to be simply a demonstration that scientists under attack usually have resources to defend themselves, fine. But as written, it seems to be just the sort of assessment of the merits of a piece of experimentation that Pinch says can't be made.

As written, it's also wrong: the fact that a dye, added *at a single point in time,* disperses evenly throughout the cell in a short time says nothing about whether that degree of stirring is adequate to ensure that the temperature is uniform throughout the cell—the real point in question—when heat is being *continuously* generated in one region and lost in others. That's not to say that the stirring *was* necessarily inadequate (I think it was, but that's not important here), just that this dye experiment is irrelevant to the issue. By making such a statement Pinch is acting inconsistently with his own precepts and doing exactly what he criticizes McKinney for.

The other example (also Pinch's) comes from a discussion of the role of rhetoric in resolving controversy (Pinch 1995). It's a very entertaining and interesting piece, but the conclusion includes the sentence "In the cold fusion case I claim that the outcome of the debate has to a large part been shaped by particular rhetorical performances." How can he make such a claim?

Of course, we need to do some second-order rhetorical analysis on the claim itself. I assume he does *not* believe that we would all now be powering our homes with little cold fusion generators if only Pons and Fleischmann had been a little more rhetorically adept. Does he merely mean that the precise course of the verbal debate would have differed in some details, which is obviously but trivially true? Or that the outcome would have been different in some *substantial* aspects—for example, that Congress would have given the multimillion-dollar award requested to keep us ahead of the competition, or that there would be centers for cold fusion research still thriving in the United States? I can't say that's impossible (although I don't believe it for an instant), but to make the claim implies a judgment of the *relative* importance of the causative forces at work, and if some of those are not to be examined, on what basis can one judge?

These are minor examples, but I think they are representative and support Bricmont and Sokal's complaint: if science studiers want to insist on agnosticism with regard to the merits of scientists' truth claims, they ought to be more circumspect about some of their own more sweeping claims when the latter imply some elements of evaluation of the very issues they've said they can't evaluate. As a certain philosopher not entirely unknown to the field once remarked, "Whereof one cannot speak, thereof one must be silent."

Chapter 29

CAUSALITY, GRAMMAR, AND WORKING PHILOSOPHIES: SOME FINAL COMMENTS

Michael Lynch

I am encouraged by the second round of papers in this volume. Most of the sociologists and historians agree that their studies of historical and contemporary cases must take into account what scientists say about their own theories and practices. It also seems that many of the scientists recognize that studies of what historical and contemporary scientists *do* will not always vindicate what scientists and philosophers *say* about science in general. I also think it is clear that what seems to be emerging is not a truce between two discrete sides. Instead, it involves a much more varied collection of agreements and disagreements that crisscross the two-cultures divide. Many substantive disagreements remain, for example, about the relationship between methodological and philosophical relativism, but the overall tenor of many of the arguments is inquisitive rather than inquisitorial. Despite my support for the overall dialogue, I would like to express a couple of reservations about some of the terms of the peace agreement that may be emerging from this volume.

The first reservation has to do with the range of views included in the volume. Physics and SSK are heavily represented, while other branches of natural science and other constituencies in social and cultural studies are not much in evidence. The predominance of physicists may encourage the antiquated view that practitioners of the "purest" science are natural spokespersons for science in general. While arguments about the reality of subatomic particles and the verification of Einstein's theories can be engrossing and engaging, we should not forget that much, indeed most, of the critical work in social and cultural studies of science focuses on particular historical and contemporary developments in biology, biomedicine, and biotechnology. There is also a strong Anglo-American presence among the sociologists and historians writing in this volume. I mention this because I am sure that it will not be lost on many readers that few if any spokespersons for Sokal and Bricmont's (1998) Francophone, Francophile, and feminist targets are found in this volume, and nobody is taking up the banner of "postmodernism." At this point, I think that most, if not all, of the scientists have acknowledged that those of us on the "sociology" side do not march in lockstep with one another, and we do not all espouse a radically relativist "postmodern" antipathy to science and rationality. Although I am happy to promote such recognition, at the same time I hope that it will not encourage the conclusion that glib caricatures of "cultural constructionism" and "postmodernism" accurately portray the views of our colleagues in Paris or Santa Cruz.[1]

There is a real dilemma here. Gross and Levitt (1994), for example, acknowledge that there are major differences between the various sociological, cultural studies, and deep-green ecologist groups they lump together under the cultural constructionist label, just as Gerald Holton (1993) recognizes that there are major differences between SSK and the more virulent antimodernist movements he lumps together under "antiscience." Nevertheless, these writers lump "us" (social historians and ethnographers of science) together with "them" (rabid and vapid antiscientists). Gross and Levitt's argument is that "we" go out of our ways *not* to denounce "them," and thus we fail to police our sector of the academy from irrational and irresponsible elements.

There are at least three reasons why many of us are reluctant to take up Gross and Levitt's invitation to stand up and denounce the loonies in our midst.[2] First, it smacks of McCarthyism, though I agree with Steven Shapin [23] that we should not delude ourselves into thinking that there are huge political stakes in this game. Second, when given a chance to speak for themselves, many of the contributors to the *Social Text* issue that Sokal so famously "hoaxed" demonstrated an ability to argue rea-

1. See Latour 1999a for an interesting defense of the "Paris" version of science studies and some of the contributions to Ross 1996a for a defense of the "Santa Cruz" version.

2. Not everyone is so reluctant. As mentioned in my initial paper in this volume, some of the most hostile and crude denunciations of social and cultural studies in Gross, Levitt, and Lewis's *The Flight from Science and Reason* (1996) are written by social scientists and philosophers.

sonably, respectfully, and knowledgeably about the sciences they studied (and, in some cases, the sciences they practiced).[3] Unfortunately, their voices were drowned out by the clamor over Sokal. Third, I know from having been involved in science studies for many years that hostile categorical judgments about radical feminists, relativists, postmodernists, and various other "ists" and "isms" are liable to sweep aside serious, provocative, and interesting research along with half-baked, outrageous, and flippant polemics. It often is necessary, for example, when reviewing articles submitted to science studies journals, to draw a line between serious work and rubbish, but I know all too well that colleagues and students whose judgments I respect often draw the line much differently than I do. Rather than go into a more elaborate discussion of the question of where to draw the line between responsible and irresponsible criticism, and between well-documented research and writing from beyond the lunatic fringe, I will simply register uneasiness about the possibility that caricatures about irrational, hostile, and ignorant versions of science, which contributors to this volume may agree do not apply to the present company, will be presumed to portray others who are not represented here.

A second reservation concerns a prevailing picture of what social and cultural studies of science set out to do. Bricmont and Sokal characterize SSK as an "intellectual movement that claims to give a causal explanation of the content of scientific theories" [14]. As Saulson points out in his commentary [22], Bricmont and Sokal base their characterizations of SSK on a reading of David Bloor's (1991) programmatic arguments. Saulson goes on to say (characterizing Bricmont and Sokal's view) that SSK "is built upon the premise that sociological factors determine the actions and beliefs of scientists." It is worth pointing out, as Saulson does, that Bloor emphasizes that "social causes" are not exclusively responsible for scientists' beliefs. Accordingly, it might seem that the main question to resolve at this point is how "social factors" (a covering term for a range of personal, circumstantial, and institutional considerations) interact with "scientific factors" (often treated as synonymous with objective reality, nature itself, or properties of the physical world) to *cause* scientific beliefs. Presumably, hostilities will end if sociologists and scientists can manage to agree about the proper balance of social and natural causes of scientific belief. My reservation has to do with the very idea that SSK should, or even *could,* show that natural or social "factors" *cause*

3. The collection of papers in the, now infamous, special issue of *Social Text* were republished, without Sokal and with some additional papers, in Ross 1996a.

scientific belief. Although this idea may be consistent with Bloor's proposals, which are often taken to be foundational for SSK, many studies in the broader field of social and cultural studies of science do not set out to give causal explanations.[4] For many of us, the task is not to decide whether natural or social "factors," exclusively or nonexclusively, determine the contents of science. Instead, the task is to describe how conceptions of objectivity, subjectivity, society, and nature are featured in the contents of science. Distinctions between social and natural factors, and between knowledge and belief,[5] are part of the *grammar* of the sciences: scientists use historically changing vocabularies of objectivity and subjectivity when they argue with one another, and they employ those vocabularies when they investigate noisy fields of data in which "social factors" have not yet been discriminated from "scientific factors." As Peter Winch (1958) pointed out more than forty years ago, the analysis of concepts and distinctions is the proper job of philosophy and not empirical social science. However, much of the best work in SSK has shown that it is possible to use case studies to illuminate the historically, practically, and interactionally situated uses of fundamental scientific concepts and distinctions. What Peter Dear calls "epistemography"—historical and ethnographic case studies that examine the contextual uses of epistemological concepts and distinctions—may help to illuminate and explicate the contents of science, but it will not necessarily yield causal propositions neatly packaged into social and scientific factors.

ROCKS, LAWS, AND WORKING PHILOSOPHIES

Having stated my reservations, I will now pursue what may be an emergent point of agreement with Stephen Weinberg. In an attempt to clarify his comparison between laws of physics and rocks in a field, Weinberg says that he was not making "a philosophical argument, but rather a personal report" on his experience as a physicist. He then concludes by saying: "This raises a larger issue. As shown by our common use of words like 'real' and 'true,' we all adopt a working philosophy in our everyday lives that can be called naïve realism. As far as I know, no one has shown why we should abandon naïve realism when talking of the history and

4. Bloor and I debate questions about the Strong Programme's emphasis on causality in an exchange in Pickering 1992. Also see Lynch 1993, chapters 3 and 4.

5. For a critical discussion of the knowledge-belief distinction in SSK, see Coulter 1989.

sociology of science. Philosophers may be able to help us to sharpen the way we understand words like 'real' and 'true' and 'cause,' but they have no business telling us not to use them" [24]. This passage gives me reason to reconsider a statement made by philosopher Ian Hacking that I quoted approvingly in my second-round contribution: "the science wars are founded upon, among other things of a more political and social nature, profound and ancient philosophical disputes" (1999, 68). I take it that Weinberg is not saying that his "naive realism" is a crude and uncultivated philosophy, but that it is a "working philosophy" rooted in his experience as a physicist. It is possible to draw some fairly close connections between what Weinberg says about the reality of physical laws and some ancient (and not so ancient) philosophical arguments, but as I understand him, he is not conceding authority to the philosophical tradition. Contrary to Hacking, he is not suggesting that his general ideas about science and his critical views of the Kuhnian legacy are *founded* on ancient philosophical positions. Instead, he is claiming that these views are rooted in the scientific life. I believe Weinberg is right and that general convictions that arise from a scientific life cannot simply be tagged to philosophical positions, ancient or modern. However, I also believe that we ought to explore the origins and limits of his "working philosophy."

A first question to explore has to do with where Weinberg's "naïve realism" comes from. He characterizes it as "a working philosophy in our everyday lives," but whose "everyday lives" does he have in mind? I imagine that Weinberg would agree that the naïve realism of a distinguished physicist in the early twenty-first century is not identical with the naïve realism of his early-modern predecessors or his nonphysicist contemporaries. The difference is not a matter of philosophical sophistication, but of historical and practical situation. The professional physicist's everyday life includes a familiarity with advanced mathematics, sophisticated instruments, and descriptive terminology that the rest of us find extremely remote from our everyday lives. This, of course, is a source of tremendous authority for Weinberg, and he and other physicists have excellent reason to insist that those of us who have spent all, or the greater part, of our professional lives in humanities and social science departments should take care not presume to investigate and criticize the "culture" of specialized physics communities until we have given serious consideration to the technical competencies that take many years for aspiring physicists to master. The same goes for other cultural activities in which knowledge is based upon cultivated skills and specialized understandings. One of the contentions of "relativist" sociology of knowledge is that a sociologist or anthropologist who aims to study the working knowledge in a specialized community should maintain a delicate combination of respect and distance. This methodological requirement applies regardless of the initial authority and credibility assigned to the community in question. In principle, it applies to investigations of astrology as well as astrophysics, histories of phrenology as well as neurology, and ethnographies of high-energy shamanism as well as high-energy physics. I tend to believe that the entry requirements for astrophysics are far more demanding than they are for astrology, and, contrary to some of my colleagues, I do not believe that the initial methodological orientation should forecast a particular form of causal explanation to be developed at the end of the investigation. However, while we may assume that a physicist's everyday life is out of the ordinary in many respects, Weinberg also ties his "working philosophy" to broader, more ordinary domains of life and language. He suggests continuity between a physicist's use of words like "true" and "real" and everyday uses of those terms, and he also trades in homely, broadly familiar examples. (See Jay Labinger's second-round contribution [18] for some critical remarks about such examples.) One needn't be a professor of philosophy in order to use terms like "truth" and "reality" in a logical and grammatical way, or even to reflect upon the implications of their use. A philosophical education is valuable in its own right, but our mastery of ordinary words is grounded in our socialization into a language.

This takes me to a second question about Weinberg's working philosophy. Even if we grant that his realism is not a dim echo of an ancient philosophical position, and that it emerges from an unspecified mix of everyday and professional experience, must we suppose that it the only reasonable working philosophy for a physicist to hold? What difference would it make if a physicist espoused a radically different working philosophy, or, as Shapin's list of quotations [8] indicates, if physicists casually expressed diverse, logically inconsistent opinions about science in general? I think Weinberg would agree that his working philosophy does not follow inevitably from an everyday life in physics. He emphasizes that his arguments about laws of physics and rocks in the field are "a personal report, that my experience of the laws of nature in my work as a physicist has the same qualities that in the case of rocks make me say that rocks are real." I completely agree with Weinberg's final line: "Philosophers may be able to help us to sharpen the way we understand words like 'real' and 'true' and 'cause,' but they have no business telling us not to use them." Indeed, as I understand the lesson from Ludwig Wittgenstein's later philosophy (discussed in Trevor Pinch's first essay in this volume [2]), when *philosophers* use words like "real" and "true" and "cause," they often owe an unacknowledged debt to *ordinary* uses of those terms. Arguments about "natural" versus "social" causes of scientific belief may distract us from recognizing the extent to which social-historical and ethnographic studies of science are *conceptual* investigations. Whether or not such investigations lead to causal explanations is an open question, but such studies may illuminate what Weinberg, other scientists, and their historical predecessors are *saying* when they use the terms "truth," "reality," "objectivity," "experience," and other items in the general science glossary.[6] Whether or not we believe that currently accepted laws of physics are socially relative, we may be led to appreciate that physicists, like the rest of us, deploy the grammar of general science in historically variable, pragmatic, and contentious ways. In my view, the sociologist's job is not to combat the physicist's working philosophy with an alternative "social" philosophy, but instead to investigate its place in the collective production of science.

6. For examples of historical studies in which such conceptual themes are prominent, see Shapin 1994 and Dear 1995. For critical discussions of conceptual issues in the human sciences, see Button 1991.

Chapter 30

READINGS AND MISREADINGS

N. David Mermin

First, astrology. Even after reading my account of the exchange with Barry Barnes over their remarks on astrology, Trevor Pinch [21] continues to see me pretty much as Barnes did, "exercised by Barnes, Bloor, and Henry daring to address the matter of astrology in the same vein as they treat other more acceptable areas of science." Years of public attacks on astrology by eminent scientists seem to have created the presumption that no scientist can address the subject in any context without deploring and condemning. I did not object to Barnes, Bloor, and Henry applying to the practice of astrology the same investigative tools they bring to bear on conventional science. My criticism was that astrology provided a bad textbook illustration of an area that might shift across the boundary scientists draw between science and pseudoscience, because far too much scientific knowledge would have to be thrown out to accommodate astrology with any plausibility.

This same disposition, to read what he expected me to be saying rather than what I actually did say, appeared in David Bloor's reply to my review of *Scientific Knowledge*. In criticizing their choice of astrology as an example I emphasized that there was a difference between drawing sharp, precisely defined boundaries between science and nonscience and distinguishing between extreme cases: "Boundaries are indeed hard to draw; I would have said there is very little point, and much potential mischief, in trying to draw them at all. But far away from these highly problematic borders it is no harder to identify extreme types than it is to agree that a cubic mile of sand is a heap, while ten grains are not a heap [even though it is impossible to specify a precise number of grains beyond which a non-heap becomes a heap]" (1998a, 622).

Bloor pounced on the first of these two sentences, charging me with "exercising his right to sustain the boundary around science by quick-fire judgments, while covering up his tracks by declaring his indifference to the entire question" (1998, 626). But my second sentence about gross distinctions, which Bloor ignored, is entirely consistent with my remarks about the inappropriateness of astrology as an example. The sentence he did address is only about the futility of trying to set up rules that will enable one to make arbitrarily fine distinctions.

It's important to get past this stage of trivial misunderstanding so we can get to an agreed-upon residue of genuine disagreement. In the case of astrology, the real question is whether some interesting, nontrivial issues might not be swept aside by renouncing a priori the possibility of any sociological basis for a distinction between science and superstition. To categorically declare such a question beyond the scope of any proper sociological investigation seems to me as arbitrarily restrictive as the categorical ban on any use whatever of present knowledge in investigating the past. I'm inclined to agree with Peter Saulson [22], that establishing such absolute methodological taboos is a form of the time-honored, but not invariably wise, scientific practice of putting on the shelf problems that seem too hard to solve.

Turning from astral destinies to predestination, I like Peter Dear's theological analogy [16]. But isn't Steven Weinberg's final theory a little closer to salvation than my overworked tapestry? The tapestry is more of a process than an end result. I didn't mean to suggest that it could never be unraveled and rewoven—just that this would be a far from easy undertaking, not merely a matter of snapping a thread here and splicing together a couple of loose ends there. In 1933 it was surely much more efficient to assume that Miller screwed up in one way or another than to attempt the massive unraveling and reweaving of the contemporary form of the relativity tapestry necessary to make sense of a positive result. By all means encourage scholars to investigate scientific practice as a social activity, but don't do it in a way that misleads them into thinking that putting together convincingly all the pieces of a scientific puzzle is a more flexible process than it actually is.

By setting my disagreements with Harry Collins and Trevor Pinch on almost as high a metaphysical plane, Michael Lynch [19] risks losing sight of my initial criticism. I was bothered by the distorted picture of scientific knowledge offered the general reader—much the same complaint that people in science studies direct at scientists, often with good cause. It seemed to me that a book claiming to tell the general reader "how science really works" ought to have emphasized, as an important part of the presumption that Miller's results could not be correct, what many contemporary physicists regarded as substantial independent evidence in support of relativity. In our subsequent conversations it emerged that Collins and Pinch took this additional evidence to be implicit in what they did cite—"the culture of science"—and that they are more skeptical than I am about how compelling the contemporary evidence actually may have been. I still believe such subtleties required more elaboration than they were given, even in a semipopular book, if the purpose of the book was to tell nonscientists what they "should know about science."

Descending further down the metaphysical ladder, I note that Jean Bricmont and Alan Sokal [14] end their list of what underlies the "relativist Zeitgeist" with "a lot of concrete empirical work, on which we make no judgment." But the quality of that empirical work is central to evaluating the whole undertaking. If it is set aside, one loses the most stringent test of whether "knowledge" is more productively viewed as situated in people's heads or the mirror of something objective. My own concerns, as I've tried to describe them in this volume, are that some of the attitudes toward knowledge embodied in methodological doctrines may impose an unnecessary rigidity on the concrete empirical work that results in a failure to get the content of the knowledge quite right. This danger exists whether you choose to view the knowledge as engraved in the Book of Nature or negotiated by a scientific community—a dichotomy which I'm increasingly inclined to view as yet another example of Bohr's broader notion of "complementarity."

Abandoning metaphysics entirely, I'm puzzled by Jane Gregory's declaration [17] that scientists expect certainty of themselves and educate their students into the "science-as-certainty philosophy." To be sure, there are as many silly people in science as in any other field of activity, and a few of them may well harbor such illusions. But if the rest of us have learned anything from our education and professional experience, it's that you're always dealing with probabilities. Even those of us who believe in final theories know perfectly well that the computational—indeed, the conceptual—gap between immutable law and particular phenomena can be so hard to bridge that there's plenty of room for inspired guessing, uninspired confusion, or just plain error.

Recently the difficulty in communicating across disciplinary lines was vividly brought home to me by an unpleasant experience I had while reviewing sociologist Thomas Gieryn's *Cultural Boundaries of Science* (1999) for *Nature.* Having become painfully aware of the difficulty of writing something that sounds sensible to members of one community that does not sound like blithering idiocy to the members of another, I took considerable care to describe how I, as a physicist, reacted to Gieryn's approach, in a way that I hoped would not strike a sociologist as hopelessly simpleminded and naïve. But my efforts to do this were systematically undermined by the copy editor at *Nature,* who, under the impression that he or she was clarifying my remarks, imposed on my review a point of view distinctly alien from mine and then ignored my urgent explanations of why the version in page proofs was unacceptable. Here are two characteristic differences between the review I sent to *Nature* and what they published under my name (Mermin 1999):

I began the review with two quotations I had noticed since reading the book, which struck me as two quite different examples of the kind of boundary drawing that interested Gieryn. I then said, "I have learned from Thomas Gieryn to recognize specimens like these as exercises in 'cultural cartography' or 'boundary work.'" As reworked by *Nature* this came out "Thomas Gieryn tells us that specimens like these are. . . ." I would guess this was done to make the first words following the two epigraphs refer to the author rather than the reviewer (but then why not "Thomas Gieryn has taught me to recognize . . ."?). But note the distortions introduced by this change: (1) In my version it is clearer that the specimens were found by me, not given in Gieryn's book; (2) The important statement that I learned something from Gieryn's book has been dropped; (3) My agreement with Gieryn that such statements are exercises in cultural cartography has been weakened to the assertion that according to Gieryn they are. Thus my sympathetic opening has inadvertently been transformed into something that can be read as somewhere between guarded and downright hostile.

I concluded my review with the remark that it would be a good thing "if one outcome of the science wars were to make physicists less uncomfortable with their professional deployment of rhetoric." *Nature* changed the second half of this to "less uncomfortable with using rhetoric when describing their work." Apparently the notion that the rhetoric could be an inseparable part of the work was so foreign that the copy editor felt it necessary to introduce an explicit distinction between the work and its description, which undermined the very point I was making. My putting the rhetoric on the same footing as the gathering of data and the mathematical analysis so offended editorial preconceptions that it required this kind of disfiguring "correction." As a result, *Nature*'s reworked version of my review displays just the kind of discomfort with the use of rhetoric in science that I was urging my fellow physicists to overcome.

I mention this here partly because I am still in a rage over the dirty trick *Nature* played on me and want it to be somewhere on permanent record (they refused to print an erratum) that the text they presented to their readers as mine was not. But there is also an important lesson in this unfortunate episode for those of us engaged in the science peace process: every word counts. It is all too easy to be misread and to misread others. The text of my review triggered a series of misreadings in *Nature*'s copy editor, who then froze them into the text by a series of damaging alterations. I do not attribute the action of the copy editor to malice or idiocy. I attribute it to strongly ingrained ways of thinking resulting in an inability to acknowledge that something more subtle than mere clumsiness or error might conceivably have been behind my words.

This is not just a problem for copy editors. Much of what I have to say in these pages is about misreading, both of me and by me. We should all be on guard against it.

Chapter 31

PEACE FOR WHOM AND ON WHOSE TERMS?

Trevor Pinch

Several of the authors in this volume reflect on the wider debate in which we are engaged. The science wars is indeed a peculiar debate, and it is not clear what the long-term effect of gathering together representatives of both sides in this volume will be. Will it help bring the controversy to an end? And on whose terms will it end anyway?

Most controversies in the natural sciences are brought to a close in a comparatively short period of time with a clear victor. In the case of cold fusion, for example, scarcely a few months elapsed between Pons and Fleischmann's initial discovery announcement at a press conference and the setting in of widespread disillusionment over their claims. Within a year it was all over. This does not mean to say that some scientists did not continue to fight the good fight for cold fusion long after (some still continue today), but the majority of the community closed ranks very early on. Most debates in the humanities and social sciences are seldom resolved so quickly and easily. "Free Will proclaimed as victor over Determinism at philosophy conference!" is not a headline one will find.

I am struck when I look at scientific controversies by how good the scientists are at closing them down. Why this is so is, of course, itself a matter of dispute. For those of us in science studies such debates are settled by the presence of an effective community—a core set—within which a combination of rhetorical, cognitive, material, and social resources are brought to bear. For scientists it is more the case that the "truth will out," sometimes by means of a few crucial experiments in tandem with irresistible theoretical arguments. But whatever the means of closure we can ask whether the science wars as a controversy will follow the natural science model of closure or the humanities model.

I think many scientists in their heart of hearts initially hoped and believed that the debate would be closed down in the natural science way. As Peter Saulson nicely reveals in his second-round contribution [22], upon first hearing of the Sokal hoax he felt that a blow (perhaps a mortal blow) had been struck for his side. It also seems to me that Alan Sokal, at least initially (perhaps still to this day?), operated with a model whereby a fairly definitive refutation was all that was necessary. Today proponents from both sides caution us that things are not going to be settled so easily. The irretrievable philosophical differences underpinning both sides are pointed to, differences as to how to characterize and explain complex historical and social phenomena.

The science wars as a whole are fought out within a variety of forums and according to many different sets of conventions. Within this strange hybrid forum—part cocktail party, part scholarly exercise, part diplomatic exercise—it is evident that the debate seems to be moving toward a standard humanities form of resolution. The take-home message is that the issues are difficult and subtle and there are no easy resolutions. This is perhaps not surprising since the didactic forum here is closer to the humanities than anything else. But this will not persuade the warriors on both sides: those who either seek the definitiveness of the natural science resolution to the debate or those who reject science *tout court*.

I am tempted to leave things here except that there is another arena where the science wars are deeply consequential. I refer not to jobs in science and technology studies. As Shapin notes [23], if the scientists are proclaimed victors and a few hundred academics lose their jobs or are shunned, it probably will not make that much difference in the long haul. But there is one area where most of us seem to agree science is in trouble—namely, the public's failure to understand it. It is here where I suggest the impact of the science wars is more crucial and where the stakes are somewhat higher.

As might be expected this dimension is already enlisted by both sides in the debate. Collins and I in *The Golem* and *The Golem at Large* claim that the public needs the science studies understanding of science in order to better deal with the process of science. Bricmont and Sokal claim that it is our impact which is part of the problem—we are encouraging irrationalism amongst the public. But the problem goes deeper than this, as David Mermin points out. The failure of the public to deal with science goes back a very long way, long before social constructivism had its moment in the sun.

Of course, not everyone working on the public understanding of science is agreed as to exactly how the science studies model will play out. For instance, Jane Gregory [17] takes issue with me over how to view this wider public and its understanding of science. It is a public which she says has already been "thumped and coddled" by the golem at large and thus does not expect much of its scientific experts. Collins and I, on the other hand, think that the public still pays homage to the standard model of the relationship between scientists and their public. But even if Gregory is right, it surely only adds fodder to the argument that there is much to learn from the science studies model. If the public really are as savvy as Gregory claims, then any model of science which scientists and government bureaucrats hold that is so out of kilter with the predominant way of thinking will likely breed trouble in the long run for science.

This is the area where the science wars matter. Different models of science and its relationship to the public are on offer. One model holds that scientific truths are certain truths and are very different from anything else in society. Within this model (which it seems is held by very few people in this volume) there is a firm boundary between scientists and the rest of society. The more contingent model, which has scientific truths woven from strands into rope or into tapestries, offers a view of science which commands respect but which sees scientific expertise as more like other communities of expertise. Which model wins respect amongst the public in the long run is an issue that is too important to be left to the outcome of this arcane science war.

Chapter 32

PILGRIMS' PROGRESS

Peter R. Saulson

After a brief comment on Steven Shapin's round 2 essay, I devote the rest of my remarks to a discussion of some of the ideas presented by Steven Weinberg.

FEARFUL SYMMETRY

One of the great pleasures in the earlier rounds of essays in this volume was reading and pondering the wonderful set of "antiscience" quotes collected by Steven Shapin [8]. He taught some valuable lessons about the nature of "antiscientific" ideas, as well as about how our reaction to an idea can depend on whether it comes from an orthodox or heterodox source.

I hope he will not take it amiss if in a similar spirit I ask readers to think about an imaginary case of someone else uttering a sentence from his own second essay [23]. I was struck in particular by one sentence that appeared to have even more resonance than he intended. He asked us to consider the Pyrrhic character of a hypothetical victory of the Defenders of Science, which if it occurred would mean that "members of one discipline will be licensed not just to evaluate the work of another in which they are not trained, but they will be authorized to characterize its points and motives in ways that members of the indicted discipline systematically reject."

That sentence brought my reading abruptly to a halt. It sounded just like something that I wished I had said once. But in that cranky moment earlier in my life, before I had spent time discussing these matters over a beer (as Shapin suggests in his first essay), I had wanted to utter similar words to complain, not about victorious Defenders of Science, but about the impertinence of the practitioners of science studies. Doesn't the very existence of a field like the social study of science depend upon the work of people trained not in science but in sociology, history, or whatever? Their discipline calls them to evaluate (in some sense of the word) the work of scientists. And much of the fun of science studies comes from characterizing the points and motives of science, usually in ways that most scientists would systematically reject.

I wouldn't want to insist too rigidly on the exactness of the suggested symmetry between our two communities. As Shapin points out, they differ in size and social position as much as they do in interests. Instead, I simply want to argue in another way for what Shapin called for in his first essay, that we find every possible way to seek out our common interests.

TRUE, MOSTLY

Steven Weinberg offered some provocative remarks in his previous essays. Perhaps the most striking was that, to him, the laws of physics are "as real as rocks" [9]. As a physicist myself, I know what he means, even though his idea may seem strange to others. We physicists are brought up to believe that discovery of Nature's laws is our object and that our profession has already achieved a large measure of success. Indeed, we spend most of our education learning the laws of physics that have already been discovered. To the extent that we already possess in written form many of the laws of physics, it would perhaps not be surprising to accept their reality, even if they can't stub our toes.

Michael Lynch [19] questions the attribution of reality to the laws of physics on the grounds (among others) that "facts are not things, pure and simple, separate from language, and while laws may describe natural regularities, formally they are less like rocks and more like rules, propositions, and maxims." Experimental physicists, and theorists less convinced of their direct line to Nature than is Weinberg, are quite likely to share Lynch's belief. Even Weinberg confesses [24] that he uses the concept of reality mainly as a term of respect.

Whatever the outcome of this argument, I want to point out that Weinberg's essays make an important and unusual claim, even if we translate his statement to the less outrageous-sounding "The laws of physics are true."

First, it is important to note that Weinberg carefully distinguishes between "hard" and "soft" aspects of the laws of physics, in the context of evolving knowledge. The hard aspects are those which will survive the changes, the soft those which will be discarded. Already, this is problematic. How is one, without the benefit of hindsight, to know which ideas will survive future discoveries? It gets even worse. Weinberg makes it clear that he believes that we are in search of a "final theory" of physics, from which all other ideas in physics will be derived. In other words, *none* of our present theories of physics are strictly correct, but are only approximations to the ultimate truth.

American listeners to public radio's *A Prairie Home Companion* will be familiar with host Garrison Keillor's praise for his supposed sponsor Powdermilk Biscuits, which he praises as "pure, mostly." We seem to be in a similar situation here, with laws of physics that are "true, mostly." If all of the laws of physics have this character (except the postulated final theory), how can it be that the laws have the kind of truth that would motivate us to describe them as being "as real as rocks"?

Of course, Weinberg believes that among our present state of knowledge there exists a large measure of "hard" physics, on which we might agree even with creatures from other planets. Within restricted ranges of applicability, we have laws that describe outstandingly well the behavior of the natural world. Newton's laws will always be "true, mostly" in this sense, as will Maxwell's equations of the electromagnetic field, Einstein's field equations for gravity, and so on.

Nonphysicists need to be reminded that what most interests most of them is probably among the "soft" material destined to be swept away. The "vision of reality that we use to explain to ourselves why the equations work" includes nearly all of the metaphysical concepts that accompany our theories: Newton's absolute space or Einstein's elastic spacetime, the luminiferous ether or fields in the vacuum, the deterministic trajectories of classical physics or the probabilistic wave functions of quantum mechanics. Essentially all of the popular science literature is thus destined for the dustbin, since that genre slights the laws themselves in favor of explication of what they "mean." It remains to be seen what will remain when (or if) we should finally arrive at a final theory. But none (or almost none) of those grand philosophical concepts partake of the reality which Weinberg attributes to the laws themselves.

This is weird stuff. Weinberg is defending the absolute character of approximate truth. Unusual as this may sound, it is based on ideas that are gaining a strong foothold in physics. On a variety of levels, the more complete laws do appear to allow very particular natural approximations, which although they have less generality usually have greater simplicity. (These are the so-called effective field theories.) Something wonderful is going on—a well-studied regularity of the world, encapsulated in a simple law, keeps its claim to (limited) validity even as our knowledge is extended, by virtue of this kind of hereditary relationship of theories. That physical laws can have this kind of relationship to one another is starting to appear as one of our deepest ideas about Nature.

The historical process of discovery has usually found the more limited, simpler law first, and only later the more general law that includes it as a specific case. The hope that history may repeat itself once more is behind the general optimism of theoretical physicists and specifically motivates those who, like Weinberg, hope that we are near the discovery of a "final theory" that will incorporate and extend all of the laws that we know so far.

ARROWS AND TARGETS

For Weinberg, the distinction between "hard" and "soft" knowledge is whether (or to what extent) the laws are "culture-free." He uses a metallurgical metaphor to describe the refining process by which science separates out the cultural "slag" from the pure element of truth. Mathematical descriptions of the world persist (perhaps by incorporation in more general formulations) while the culture-laden concepts that attended their birth are discarded as new knowledge is acquired. So the distinction between hard and soft knowledge is inextricably linked to a notion of scientific progress.

Weinberg goes out of his way to contrast his point of view with Thomas Kuhn's position that it was nonsensical to talk about scientific progress as taking us "closer to the truth." Weinberg's defense of his position is an insider's, convincing mainly to one who understands physics (or who trusts those who do): our present ideas really are better ideas, in that we have a more complete and more coherent understanding of the world. While I share this belief, it is an inescapably "cultural" claim, one that lives in the world of shared ideas.

I think Weinberg could add another line of argument that would both strengthen his case and make clearer what he means by "culture-free" knowledge. I propose the following thought experiment: ask anyone with a position of responsibility in our industrial economy whether she would care to do her job using only technology uninfluenced by scientific discoveries of the last seventy-five years. My prediction is that you will get few takers.

What does this prove? Only the obvious point that there is a measure of progress in science that has nothing to do with coherence or understanding or anything else that lives in the world of ideas, but instead has to do with the power to do things that ordinary people would like to get done. In this sense, progress in science is unlike biological evolution; the replacement of dinosaurs by mammals may have been strictly contingent, but our present-day knowledge is undoubtedly more *powerful* than the knowledge of the past.

Thinking about science in this way also lets us think about how the choice between two competing scientific ideas can be seen to be "culture-free." Weinberg's industrial metaphor of "refining" away the cultural "slag" isn't just a metaphor. Choices made in the industrial world are part of the way in which our social world validates knowledge of the natural world. A theory that is more powerful (in the sense of describing the world better) will be preferred by those with absolutely no cultural commitment to the content, expression, or metaphysical baggage of the theory. If it helps them to do their job, in a world in which getting the right answer counts, that theory will have powerful allies.

Few scientists would be willing to suspend their own judgment on the validity of a scientific theory until it was adopted by factory managers, but the existence of this social process is reassuring nevertheless. It happens over and over again, in unlikely ways. For many decades, for example, Einstein's general theory of relativity was accepted by scientists as an improvement over Newton's theory of gravity, even though no one could imagine any terrestrial situation in which their predictions would differ enough to matter. But precisely such a difference now plays an essential role in a multibillion-dollar industry. The global positioning system (GPS) has become ubiquitous in the military world for which it was designed as well as in such benign activities as hiking in wilderness areas. The precise time keeping on which GPS depends is sensitive to a tiny relativistic effect, that the rate at which a clock registers the passage of time depends upon the gravitational field in which the clock sits. The orbiting clocks on the GPS satellites wouldn't give the answers on which the system depends unless their special gravitational environment were taken into account. This aspect of general relativity now has a huge new world of allies, not to any of the "soft" aspects of the theory (most hikers don't care whether or not gravitation and inertia are strictly equivalent), but to its ability to accurately describe the world. If you equipped a GPS receiver with a switch that allowed you to use Newton's law instead of Einstein's, no sane person would set the switch to "Newton."

The industrial test sometimes argues that a new theory does not represent progress. The cold fusion episode gives an example. At the height of the controversy, it was suggested by some that opposition to cold fusion from "establishment" scientists was due simply to their fear of losing resources. This view was especially popular in Utah, home of one of cold fusion's proponents. If the controversy had to be resolved only at the cultural level, who was to judge between some people's commitment to a particular understanding of nuclear physics and others' belief that the state of Utah had a key role to play in human progress? But no matter what social power was held by mainstream scientists, there was a way the cold fusion side could triumph indisputably: start to sell cheap energy. Anyone would buy it and in so doing would buy into a new scientific belief. That no such event has occurred tends to suggest that the cold fusion claims were mistaken. (Of course, this could change tomorrow, if some new development is announced.)

The industrial test is certainly not perfect. This particular episode can't be understood unless one also raises questions like whether enough capital was invested in exploring cold fusion, or why we believe in the nuclear physics behind mainstream fusion research in spite of the fact that it is still far from success after many decades of work. The only point that seems certain is that it demonstrates that society offers another path for an idea to succeed outside of the ordinary "cultural" channels.

I have been talking about an idea of scientific progress that is about replacing old theories with ones that are "more true." This says nothing about what is really the fundamental disagreement between Weinberg and Kuhn, the question of whether there exists an absolutely true "final theory" of physics. All that I claim is that we can find an arrow of progress, not that we can know that the arrow will strike a target. Perhaps Weinberg's faith will be justified by future events. Until then, it remains simply a scientist's millennial dream.

Chapter 33

HISTORIOGRAPHICAL USES OF SCIENTIFIC KNOWLEDGE

Steven Weinberg

First, a response to Peter Dear's remarks on Whig history [16]. Did I really give the impression that I think the ideas of scientists today somehow cause the work of scientists in the past? Of course the work of scientists in any era is motivated by the data and the calculations available to them, as well as a whole host of social, cultural philosophical, and psychological influences. But the data are the way they are (at least some of the time) because that is the way the world is, and we know more now about the way the world is than scientists did in the past. So why not use this knowledge? I admit that there would be no need to do so if we had perfect knowledge of the data and calculations available to past scientists, but this we never have. As my J. J. Thomson story [9] was meant to illustrate, knowing something more about the world than what the scientists of the past knew can help us to fill in the gaps that will always be in the historical record.

Apropos of J. J. Thomson, I have to protest to Trevor Pinch [21] that I did not mean to be uncharitable to the discoverer of the electron. My point was not that he was careless in the experimental runs that gave values for the mass to charge ratio of the electron that were furthest from the modern value, but rather that it seemed unlikely that Thomson was more than usually careful in those runs. Harry Collins is correct [15] that Thomson might have thought he was being particularly careful when in fact he was being particularly careless, and vice versa, and this would be an interesting alternative explanation of his preference for certain values of the mass to charge ratio, but how would Collins or any other historian be able to make such a guess if he did not know the modern value of this ratio?

Knowing the truth can also lend interest to the story. Would we really care so much about the historical records of medieval Greenlanders lost in the Atlantic if we did not know (as they did not) that North America was out there to the west? I agree with Collins that we don't always have this advantage—certainly not when we tell the story of contemporary scientific research—and this does not necessarily make such history boring. But if Collins agrees that the history of a battle gains interest from a knowledge of whether the generals of the time based their decisions on correct or incorrect information, then why does he not see that the history of a piece of scientific work gains interest from a knowledge of whether or not the work was based on correct or incorrect data or calculations?

On the analogy of military and scientific history, I find Collins's remarks revealing. I am not going to argue with his distinction between the two kinds of history, that after a war the "locus of authority" moves from the generals to the historians, while after a scientific discovery it remains with the scientists. I doubt if this really makes it less clear what the role of the professional historian of science should be. The history told by scientists can be just as self-serving as the history told by generals. But even if the use of present scientific knowledge would put the historian's rice bowl at risk, surely this is no reason not to use all the information we have in exploring the past.

Chapter 34

BEYOND SOCIAL CONSTRUCTION

Kenneth G. Wilson and Constance K. Barsky

Our first topic in this third round is a response to comments by Trevor Pinch on our own first-round chapter. Our second topic is what we believe to be at the heart of the controversy the other authors are engaged in, and our views on this controversy.

In our first chapter in this volume [11], we proposed that the increasing accuracy of observations and of predictions of planetary motions presents a puzzle for sociologists. We suggested that they should find the centuries-long persistence of increasing accuracy to be incomprehensible, given the growing complexity of the organizations responsible for producing the observations and predictions. According to the expectations of many sociologists, organizational dysfunctions (such as those leading to the initial flaws of the Hubble Telescope) should have been so pervasive as to prevent the increasing accuracies that have been reported.

In response to our claim, Trevor Pinch [21] suggested that Philip Mirowski has already addressed our concerns, but in the context of the fundamental physical constants (Mirowski 1994). These constants, including the charge of the electron, have been claimed to be improving in accuracy over the past century. The accuracy of the fundamental constants is discussed in detail in a book by Brian Petley (1985) that is cited in Steven Shapin's article [8] (and also cited by Mirowski). Unfortunately, Mirowski makes no mention or critique of the physicists' claims of increasing accuracy *over time* for the values for these constants.

A useful analysis, reported by Petley, is a calculation of the errors in the values of the fundamental physical constants as reported in 1929,

1948, and 1963. Petley determines these errors through a comparison with the results obtained in 1973, which are claimed to be considerably more reliable. The comparisons suggest that the measurements of 1963 and earlier have considerably greater errors than was claimed at the time. Two of the 1929 numbers have seven times their claimed errors, while two of the 1963 numbers have around five times their claimed errors. These discrepancies lead Petley to urge that the underestimates of errors be studied further. However, he makes no reference to possible sociological effects of the kind we discussed in our article. Mirowski does not discuss such effects either.

The article by Michael Lynch [4] does raise questions closer to the kind we discussed, but Lynch raises concerns about the air transportation system and not about the systems of organizations that have generated continuing improvements in the fundamental physical constants. As far as we know, the question of possible sociological contributions to underestimates in the errors of increasingly accurate physical measurements remains largely unexplored.

Only a tiny fraction of all scientific results have reached the accuracy now claimed for measurements of planetary motions or for the fundamental constants of physics. In both cases, the reported accuracy is sometimes less than one part in a million. This means that even if the actual errors were to be one hundred or one thousand times larger than the claimed errors, the results would still have an impressive accuracy in an absolute sense. Moreover, any truly large errors in the fundamental constants should be easily visible in thousands of applications that now rely in some way on the tables of the fundamental constants having a reasonable level of accuracy.

In contrast, for a measurement that is much newer, one that has only been repeated once or not at all, an actual error seven times the claimed error could have far more serious implications. In such circumstances, continued and more careful repetition of the scientific research in question is, we believe, both a likely next step and essential to clarify what is going on.

We have focused on examples of very accurate measurements obtained through a process of continuing improvement. We believe that a careful analysis of the history of such improvements could provide better understanding of both the magnitude of error underestimates in science and their causes. Such understanding could help people in other walks of life make judgments about claims made by scientists, whether accurate or otherwise. We also believe that the struggle to distinguish science from pseudoscience could be helped by an improved understanding of the process of continuing improvement in science. This process is a hallmark of science but not of anything that we would call pseudoscience. But, we stress, there seems to be a lack of actual research evidence on the nature and the outcomes of the process of continuing improvement in any aspect of science. Thus the main purpose of our first-round contribution was to call attention to the need for more research on this process and to urge that sociologists and scientists collaborate to carry out this research. Overall, there has been little response to our article by other authors in this volume. The cause of this, we suggest, is that our concerns are more limited than the concerns of the other authors.

We turn now to our second topic: outlining what we believe is the issue at the heart of the "science wars" controversy discussed by the other authors and explaining where we stand on this issue. The heart of the matter, we believe, is that many philosophers and sociologists want "truth" and "reality" to be exact. To them "truth" cannot have an error bar. "Reality" cannot have any residual uncertainty attached to it. But as long as the requirement of exactness is imposed on the terms "truth" and "reality," there can be no mystery left about these terms. We claim that they have always been and always will be unknowable.

Weinberg has offered the most optimistic scenario we can imagine for approaching, but not reaching, an exact truth about nature. To amplify, imagine that what Weinberg suggests could happen [9] actually does happen. Suppose that future physicists come up with a "theory of everything" which, among other accomplishments, contains potentially exact predictions for the values of the fundamental physical constants. Then imagine that for centuries afterward, the measured values of these constants come closer and closer to the theoretical predictions as the errors on these measurements and on computed theoretical values both decrease with time. Unfortunately, it would still be impossible to claim that no inexplicable anomaly could emerge at even higher levels of accuracy. A precedent for such an occurrence was set by the very small anomaly in the perihelion of Mercury that could not be explained using Newton's laws at the time it appeared.

Our conclusion, we believe, applies also to results that many scientists already consider exact. One example is the claim that all electrons have identically the same mass. Brian Petley has already noted that this claim is "an act of faith," in words already quoted in Steven Shapin's article. We agree that it is an act of faith, if the claim is for exact identity. Our reasoning is that if more accurate measurements of the electron's mass start generating multiple values for the mass, and continuing and more accurate repetitions of these measurements establish that the range of values cannot be argued with, it would be impossible for scientists to do anything other than acquiesce to the results and try to understand them.

What has startled us is a claim made by Thomas Kuhn (1996), namely, that when scientists are presented with anomalies seemingly just as inexplicable as a minute variation in the electron's mass would be, scientists find it relatively easy to come up with initial, tentative explanations. What might a first hypothesis be that could explain multiple values of the electron mass, within a very tiny range, should such values be found in future experiments?

We suggest that one hypothesis might be that a new species of elementary particle exists with a mass and charge so small that such particles, by hypothesis, are rarely generated or present in today's experiments. However, when these species are present they could form bound states with an electron that would then have a mass and charge very slightly different than for an electron in isolation. The anomalous experiments would be explained, tentatively, by assuming that the electrons that depart from normal are in a bound state with one or more of the new elementary particles. It would be assumed that such bound states are so rare that they were not seen (or perhaps, not recognized as anomalous) before the anomaly became known.

This example illustrates that sometimes it may be easy for scientists to come up with an initial hypothesis when faced with anomaly. However, it is often much harder to arrive at an explanation that survives as measurements continue to improve. The development of quantum mechanics and the current theory of quarks are two examples of hypotheses that took time and effort to develop. But we suggest that Kuhn may be right in his claim that initial hypotheses are easy to produce more often than one might expect.

To wrap up the discussion of our second topic, we offer a related quote from one of the very few practical books written by a scientist to help other scientists cope with the realities of scientific research (Wilson 1990). The author, E. B. Wilson Jr., was not a sociologist, but nevertheless was very aware of the extraordinary range of difficulties that can disrupt even the most well designed of experiments. He makes clear that obstacles and blunders are unavoidable in the practice of scientific research, and that it is no easy matter to arrive at reliable scientific knowledge. We offer the quote as an addendum to the list provided by Steven Shapin:

> Finally, it probably can be said that no generalization is ever completely true and that few generalizations based on very many data are completely false. . . . The facts that the sun has risen daily millions of times and that there exists a well-founded theoretical explanation of these observations do not rule out the possibility of eventual exceptions. It seems certain that there will come a day when the sun will fail to rise. Scientists, when confronted with failures of long-established generalizations, are not usually as chagrined as philosophers think they should be. They [adapt their theories] and go on making new predictions as if the old ones had not broken down. (160)

To conclude, we return to our first topic. We have focused on the process of continuing improvement in science as the driving force behind increasing accuracies in scientific measurements because we believe it generates results of great practical and fundamental importance. The process, while not error-free, can be used for many purposes by many diverse cultures as long as the measurements become accurate enough for their errors not to matter. Meanwhile, the larger body of scientific results that have yet to benefit from many rounds of continuous improvement can be problematic for any culture that tries to use them. There are many results from the natural sciences that are in this latter category; there are many results from the social sciences that are in this latter category too. We suggest that research on the nature of continuous improvement in science could be beneficial for future users of both the natural and the social sciences.

Chapter 35

CONCLUSION

Our goal for this book was to demonstrate a new style of debate between former antagonists—an appreciation of the depth and subtlety of the problems, a little convergence, and the turning of unproductive rows into productive disagreements—and thus to take the debate beyond the science wars. We believe we have achieved this. In this concluding section, we would like to sum up some of the areas of agreement and disagreement that have come through most clearly. We have recast some of the latter in the form of "open questions." Apart from sharpening future debate, we think that they would actually make rather good questions for discussion in university seminars and similar settings. Finally, we will offer some thoughts on the wider significance of the issues we have wrestled with throughout this project.

Starting with agreements, the authors appear to converge in the opinion that science studies is *not* hostile to the interests of science, either deliberately or as an unintended byproduct. Several of the first-round position papers, particularly those of Shapin [8], Pinch [2], and Labinger [13], specifically address the charges of undermining or being opposed to science and reject them. Dear's brilliant neologism "epistemography" [10] sums up the attitude in one word. Gregory and Miller [5] see no problem in the point of view of the public, while Saulson [6] reports that at least one practicing scientist who is the subject of a case study does not feel under attack. No response contradicts this view and at least one—that of Bricmont and Sokal [14]—explicitly endorses it.

A second point of agreement is that misunderstandings and misreadings have played a major role in the genesis and continuation of the science wars. This is a central point in two of David Mermin's contributions [7, 30], but many of the pieces in all rounds rehearse this theme. Indeed, concern over being misunderstood *within* the present discussion persists right through to the final round, showing how difficult—perhaps even insoluble—this problem is. It may appear strange to trumpet an agreement that there have been and continue to be misunderstandings as a point of convergence—but see how far we have come from the propaganda-like proclamations that typified the depths of the science wars! In those episodes the antagonists rarely if ever even contemplated the possibility that their (often simplistic) accounts of the other side's position could have been a misrepresentation.

A third point is that science studies are interesting and potentially useful. But consensus here is much less clear cut: the details of just *how* they may be useful are still points of contention. Generally everyone appears to agree that it is important to study the interface between science and society, although several of the scientists would also like to see more attention to improving the internal workings of science. Both Weinberg [9] and Wilson and Barsky [11, 34] take (disapproving) note of the implication drawn from Thomas Kuhn's work, that progress in science is chimerical. They call for studies aimed at understanding what scientific progress means and how it is achieved. Trevor Pinch responds that Kuhn's influence on contemporary practice in science studies is probably overrated [21]; his first essay [2] presents a helpful discussion of some of its intellectual antecedents.

Turning to disagreements, the deepest of them turn on philosophical and methodological issues. None of the sociologists departs from the conviction that they must remain resolutely agnostic with respect to the scientific "facts" of the matter in carrying out their work. All of the scientists express at least *some* degree of concern over this commitment to methodological relativism, worrying that some aspects of a given study—possibly the ones that most interest them—cannot be addressed within such a framework. For Bricmont and Sokal this is the *key* issue: whether a purely methodological form of relativism is possible and appropriate, or whether on the contrary commitment to philosophical relativism is implicit therein. The course of this debate can be clearly traced through all three rounds of their and Harry Collins's contributions, and nearly everyone at least mentions the issue in their response pieces. Philosophical issues also comprise significant portions of other initial position pieces, especially those of Weinberg [9], Lynch [4], and Dear [10].

It is interesting that an obvious and plausible resolution to this conflict—just put aside the philosophical differences and get on with the work—does not seem to be generally acceptable, at least not yet. Indeed, many of the pieces from rounds 2 and 3 worry about *excessive* agreement, feeling that the differences are potentially productive and that there is still progress to be made in exploring them. For example, in Bricmont and Sokal's third-round contribution [25; see especially note 9], they say that sociologists should refrain from studying cases unless an independent assessment of scientific evidence is accessible to them. Our attempts to explicate this point required several rounds of e-mail exchange with the authors even as this conclusion was being finalized, and it may be worth dwelling on this issue to illustrate the productive power of disagreement.

Our most important misreading—we set it out here so as to help others avoid the same mistake—was that Bricmont and Sokal were effectively ruling contemporary controversies out of bounds to sociologists. This seemed to follow from their position because clear independent assessments of the scientific evidence are not available in contemporary controversies; the scientists disagree among themselves—the definition of scientific controversy. The authors assure us, however, that what they mean to say—a corollary of their argument against methodological relativism—is that scientists' beliefs are formed, in varying proportions, out of scientific and social factors. Sometimes, an explanation of a belief that examines purely social factors might be quite a good explanation, but at other times, when the scientific component is high, sociologists must make sure they "factor in" the scientific component so as not to give too great a weight to the social component. And when the proper evaluation of the scientific component is uncertain, the sociological conclusions will be correspondingly uncertain.

The sociologist would respond that a puzzle still remains. In studying a contemporary controversy according to this model two judgments have to be made: what is the relative weight of science versus sociology in the formation of beliefs; and if the scientific component is highly weighted, how is it to be evaluated properly? Given that scientists are in disagreement, it is hard to see how any of this can be accomplished.

However, during these rounds of clarifying e-mail exchanges, Bricmont and Sokal made an additional point. They have no objection to analysis that avoids "Whig history"—the inclusion of scientific conclusions reached subsequent to the episode under investigation—but they distinguish that from what they consider to be methodological relativism where, on principle, no evaluation of purely scientific reasoning (even contemporaneous) is allowed to enter the causal equation. Herein we may actually find a point of *convergence* with regard to the policy implications of the analysis of science, if not to the methodology of studies of scientific knowledge. Some sociologists, and surely most scientists, want to maintain science's input to policy-making even in politico-technical debates, such as those over global warming and genetically modified foods, where (almost) everyone agrees that the science is yet too immature to provide decisive answers. The too-rigid model of science with which we have grown up implies that if science cannot provide certainty it cannot provide anything. The problem for those who reflect upon scientific knowledge is to show why it is that even when science may be uncertain it can still provide much that is of use.

It would be of great interest to explore this in more detail, but we have to stop somewhere. What we have tried to do here is bring the state of the debate right up to the moment of writing and show how probing more deeply into the details of a disagreement may lead to an unexpected convergence—in this case, where the policy implications of the two positions are drawn out. We anticipate that such turns will be common, if the debate goes on in the manner we hope to have encouraged.

This leads us at last to our promised list of "open questions" for further discussion and debate:

1. Should the historian of science always, sometimes, or never take contemporary scientific knowledge into account when analyzing passages of scientific history?

2. Can and should the sociologist study unresolved scientific controversies? If yes, is the study of unresolved controversies flawed compared to cases where there is a well-established consensus on the scientific issues? Is the study of unresolved science important?

3. What is "philosophical relativism" and what is "methodological relativism?" Can methodological relativism stand alone or is it inevitably linked to philosophical relativism? Can methodological relativism be justified as a method?

4. What comes first, philosophy or observation? In other words, can a whole empirical subject and all its findings be declared invalid if it is based on a flawed philosophy? Is it necessary to solve philosophical problems before doing empirical work?

5. In what ways (if any) does science studies have potential practical value that extends beyond its role as a purely academic discipline? For society at large? For practicing scientists? What are the policy implications of sociology of scientific knowledge?

6. Do continuing disagreements between scientists and science studiers result from misunderstandings that could be cleared up with greater effort or from radical differences in language and worldviews?

Finally, why is any of this important? The editors will speak for themselves here, although we hope that our other contributors would find much they could agree with. We think of science as one outstandingly successful way of understanding the world rather than as a complete "worldview." We believe that science is by far the best way to get answers to a large number of questions—but they are not the only questions, and not necessarily the most important questions. We are concerned when science is presented as a complete description of our world, exhausting the space for every other way of looking at things—artistic, spiritual, imaginative, social scientific, or whatever. We do not think the world can be "reduced" in this way.

How are we to deal with the big problems—problems that can't be solved without science, but that science may be incapable of solving given limited time and resources—problems such as global warming, the safety of genetically modified organisms, whether eating British beef can cause Creuzfeld-Jacob disease, and so on? No matter what one's position is on what constitutes a scientific "fact," it is clear that political decisions must often be made long before the scientific facts have been established. For this reason we think it is vital for ordinary people to understand something of the processes by which scientific knowledge is made, so that they will also understand that most of what we need to know in order to make completely sound technological policy decisions is frequently unknown. Ordinary people need to understand that the mistakes which will always be made are not necessarily the result of incompetence or irresponsibility, but of the inherent uncertainty of science when it is in its formative stages, or when it asks the kind of question which is too hard to answer. Scientists remain the foremost experts on the natural world. It is just that often that expertise can deliver only the best available advice, not truth. Yet that best possible advice remains a vital component in decision-making.

And there is an important message for scientists as well. For scientists trained according to the "textbook model" of science (which is just about all of them), the temptation is strong to divide those difficult problems into the hard-core, rigorous, scientific part, and the messy, imprecise, social-political part. We don't believe such a separation is possible; still less is it desirable. In this regard, perhaps the most important function science studies can serve for scientists is that of "defamiliarization"—to borrow a term from the literary theorists—to deflect the mind out of the deeply worn ruts of standard thinking, even if only for a little while. (See Collins [12] and Dear [27] for more on this point.)

So that is why we think this book matters: it matters because the world needs responsible critiques of science. We need to go beyond the textbook model of science; science's incompleteness and its range of applicability need to be explained and explored. We need critiques that can lead us to understand how scientific expertise can and must continue to contribute to the world we live in. But we must accept that in most areas of technological decision-making, science cannot take away the responsibility for choosing the particular balance of rewards and risks we want to live with, only contribute to the choice. To generate these critiques and analyses, scientists and others will have to talk to each other. The science wars, pitting science against all kinds of external critique, work against the construction of the more modest model needed in an age when science's products are consumed, literally and metaphorically, by us all. War engenders extremes of behavior on both sides, and that endangers the kind of science, and the kind of science studies, we need.

References

Adams, Henry. 1918. *The education of Henry Adams.* Boston: Houghton Mifflin.

Albert, Michael. 1998. Rorty the public philosopher. *Z Magazine* 11, no. 11 (November):40–44.

Aronowitz, Stanley. 1988. *Science as power: Discourse and ideology in modern society.* Minneapolis: University of Minnesota Press.

———. 1996. The politics of the science wars. *Social Text* 46/47:178–97.

Bard, Allen J. 1996. The antiscience cancer. *Chemical and Engineering News,* 22 April, 5.

Barnes, Barry. 1977. *Interests and the growth of knowledge.* London: Routledge and Kegan Paul.

———. 1992. Realism, relativism, and finitism. In *Cognitive relativism and social science,* ed. Diederick Raven, Lieteke van Vucht Tijssen, and Jan de Wolf. New Brunswick, NJ: Transaction, 131–47.

———. 1998. Oversimplification and the desire for truth: Response to Mermin. *Social Studies of Science* 28:636–40.

Barnes, Barry, and David Bloor. 1981. Relativism, rationalism, and the sociology of knowledge. In *Rationality and relativism,* ed. Martin Hollis and Steven Lukes. Oxford: Blackwell, 21–47.

Barnes, Barry, David Bloor, and John Henry. 1996. *Scientific knowledge: A sociological analysis.* Chicago: University of Chicago Press.

Bauer, Martin, and John Durant. 1997. Astrology in present day Britain: An approach from the sociology of knowledge. *Cosmos and Culture—Journal of the History of Astrology and Cultural Astronomy* 1:1–17.

Bauer, Martin, and Ingrid Schoon. 1993. Mapping variety in public understanding of science. *Public Understanding of Science* 3:141–56.

Bauer, M. W., K. Petkova, and P. Boyadjieva. 2000. Public knowledge of and attitudes to science: Alternative measures that may end the "science war." *Science, Technology, and Human Values* 25, no. 1:30–51.

Beller, Mara. 1997. Criticism and revolutions. *Science in Context* 10:13–37.

———. 1998. The Sokal hoax: At whom are we laughing? *Physics Today* 51, no. 9 (September):29–34.

Benski, Claude, et al. 1996. *The Mars effect: A French test of over 1,000 sports champions.* Amherst, NY: Prometheus.

Benveniste, J. 1988. Letter to the editor. *Nature* 334:291.

Berger, P. L. 1963. *Invitation to sociology.* Garden City, NY: Anchor Books.

Berlin, Isaiah. 1998. The divorce between the sciences and the humanities. In *The Proper Study of Mankind,* ed. Henry Hardy and Roger Hausheer. New York: Farrar, Straus, and Giroux, 326–58.

Bloor, David. 1973. Wittgenstein and Mannheim on the sociology of knowledge. *Studies in History and Philosophy of Science* 4:173–91.

———. 1983. *Wittgenstein: A social theory of knowledge.* London: Macmillan.

———. 1988. Rationalism, supernaturalism, and the sociology of knowledge. In *Scientific knowledge socialized,* ed. Imre Hronszky, Márta Fehér, and Balázs Dajka. Dordrecht: Kluwer, 59–74.

———. 1991. *Knowledge and social imagery.* 2d ed. Chicago: University of Chicago Press. Original edition, 1976.

———. 1998. Changing axes: Response to Mermin. *Social Studies of Science* 28:624–35.

Boghossian, Paul. 1998. What the Sokal hoax ought to teach us. In *A house built on sand: Exposing postmodernist myths about science,* ed. Noretta Koertge. New York: Oxford University Press, 23–31. First published in *Times Literary Supplement,* 13 December 1996, 14–15.

Bohr, Niels. 1934. *Atomic theory and the description of nature.* Cambridge: Cambridge University Press. Reprinted 1985 in Niels Bohr, *Collected works,* vol. 6. Amsterdam: North Holland.

Brannigan, Augustine. 1981. *The social basis of scientific discoveries.* Cambridge: Cambridge University Press.

Bricmont, Jean. 1997. Science studies—what's wrong? *Physics World* 10, no. 12 (December):15–16.

———. 1999. Science et religion: l'irréductible antagonisme. In *Où va Dieu?* ed. Jacques Sojcher and Antoine Pickels. Brussels: Revue de l'Université de Bruxelles 1999/1, Editions Complexe, 247–64.

———. Forthcoming. Sociology and epistemology. *Facta Philosophica* 3, no. 2. Also in *After postmodernism: An introduction to critical realism,* ed. José Lopez and Garry Potter. London: Athlone, forthcoming.

Bridgeman, P. W. 1955. *Reflections of a physicist.* 2d ed. New York: Philosophical Library.

Broch, Henri. 1992. *Au coeur de l'extraordinaire.* Bordeaux: L'Horizon Chimérique.

Brunet, Pierre. 1970. *L'introduction des théories de Newton en France au XVIIIe siècle.* Geneva: Slatkine. Original edition, Paris: A. Blanchard, 1931.

Bunge, Mario. 1996. In praise of intolerance to charlatanism in academia. In *The flight from science and reason,* ed. Paul R. Gross, Norman Levitt, and Martin W. Lewis. New York: New York Academy of Sciences, 96–115.

Butterfield, Herbert. 1951. *The Whig interpretation of history.* New York: Scribners.

Button, Graham, ed. 1991. *Ethnomethodology and the human sciences.* Cambridge: Cambridge University Press.

Chargaff, Erwin. 1963. *Essays on nucleic acids.* Amsterdam: Elsevier.

———. 1978. *Heraclitean fire: Sketches from a life before nature.* New York: Rockefeller University Press.

Cohen, I. Bernard. 1952. The education of the public in science. *Impact of Science on Society* 3:78–81.

Cole, Stephen. 1996. Voodoo sociology: Recent developments in the sociology of science. In *The flight from science and reason,* ed. Paul R. Gross, Norman Levitt, and Martin W. Lewis. New York: New York Academy of Sciences, 274–87.

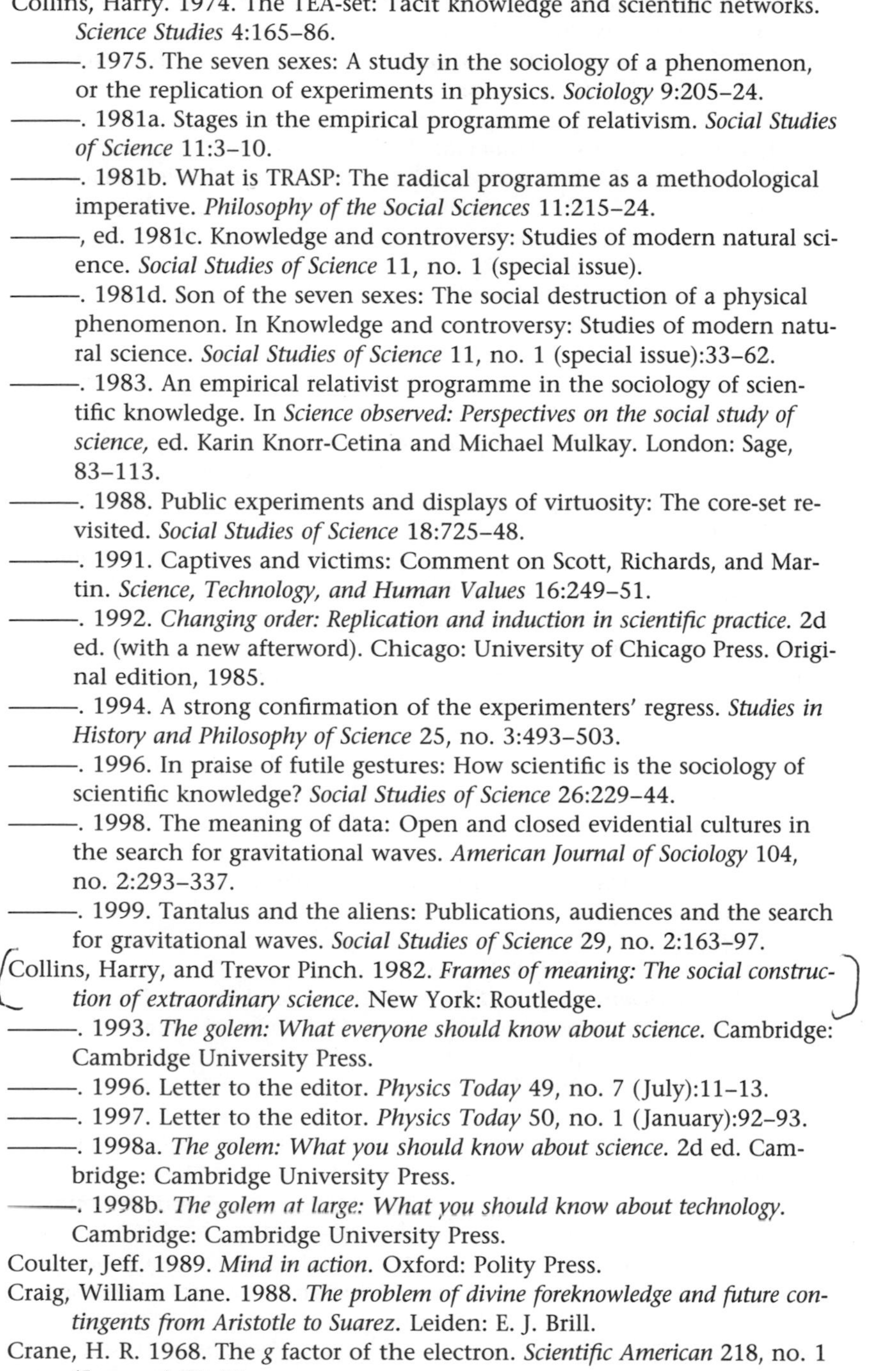

Collins, Harry. 1974. The TEA-set: Tacit knowledge and scientific networks. *Science Studies* 4:165–86.

———. 1975. The seven sexes: A study in the sociology of a phenomenon, or the replication of experiments in physics. *Sociology* 9:205–24.

———. 1981a. Stages in the empirical programme of relativism. *Social Studies of Science* 11:3–10.

———. 1981b. What is TRASP: The radical programme as a methodological imperative. *Philosophy of the Social Sciences* 11:215–24.

———, ed. 1981c. Knowledge and controversy: Studies of modern natural science. *Social Studies of Science* 11, no. 1 (special issue).

———. 1981d. Son of the seven sexes: The social destruction of a physical phenomenon. In Knowledge and controversy: Studies of modern natural science. *Social Studies of Science* 11, no. 1 (special issue):33–62.

———. 1983. An empirical relativist programme in the sociology of scientific knowledge. In *Science observed: Perspectives on the social study of science,* ed. Karin Knorr-Cetina and Michael Mulkay. London: Sage, 83–113.

———. 1988. Public experiments and displays of virtuosity: The core-set revisited. *Social Studies of Science* 18:725–48.

———. 1991. Captives and victims: Comment on Scott, Richards, and Martin. *Science, Technology, and Human Values* 16:249–51.

———. 1992. *Changing order: Replication and induction in scientific practice.* 2d ed. (with a new afterword). Chicago: University of Chicago Press. Original edition, 1985.

———. 1994. A strong confirmation of the experimenters' regress. *Studies in History and Philosophy of Science* 25, no. 3:493–503.

———. 1996. In praise of futile gestures: How scientific is the sociology of scientific knowledge? *Social Studies of Science* 26:229–44.

———. 1998. The meaning of data: Open and closed evidential cultures in the search for gravitational waves. *American Journal of Sociology* 104, no. 2:293–337.

———. 1999. Tantalus and the aliens: Publications, audiences and the search for gravitational waves. *Social Studies of Science* 29, no. 2:163–97.

Collins, Harry, and Trevor Pinch. 1982. *Frames of meaning: The social construction of extraordinary science.* New York: Routledge.

———. 1993. *The golem: What everyone should know about science.* Cambridge: Cambridge University Press.

———. 1996. Letter to the editor. *Physics Today* 49, no. 7 (July):11–13.

———. 1997. Letter to the editor. *Physics Today* 50, no. 1 (January):92–93.

———. 1998a. *The golem: What you should know about science.* 2d ed. Cambridge: Cambridge University Press.

———. 1998b. *The golem at large: What you should know about technology.* Cambridge: Cambridge University Press.

Coulter, Jeff. 1989. *Mind in action.* Oxford: Polity Press.

Craig, William Lane. 1988. *The problem of divine foreknowledge and future contingents from Aristotle to Suarez.* Leiden: E. J. Brill.

Crane, H. R. 1968. The *g* factor of the electron. *Scientific American* 218, no. 1 (January):72–85.

Cromer, Alan. 1993. *Uncommon sense: The heretical nature of science.* Oxford: Oxford University Press.

———. 1997. *Connected knowledge: Science, philosophy, and education.* New York: Oxford University Press.

Davenas, E., F. Beauvais, J. Amara, M. Oberbaum, B. Robinzon, A. Miadonna, A. Tedeschi, B. Pomeranz, P. Fortner, P. Belon, J. Sainte-Laudy, B. Poitevin, and J. Benveniste. 1988. Human basophil degranulation triggered by very dilute antiserum against IgE. *Nature* 333:816–18.

Dawkins, Richard. 1994. The moon is not a calabash. *Times Higher Education Supplement,* 30 September, 17.

———. 1996. *The Richard Dimbleby lecture.* Television broadcast. BBC-1, 12 November.

———. 1998. Postmodernism disrobed. Review of *Intellectual impostures: Postmodern philosophers' abuse of science,* by Alan Sokal and Jean Bricmont. *Nature* 394 (9 July):141–43.

Dear, Peter. 1995. *Discipline and experience: The mathematical way in the Scientific Revolution.* Chicago: University of Chicago Press.

Dewey, John. 1934. The supreme intellectual obligation. *Science Education* 18:1–4.

Dobbs, Betty Jo Teeter, and Margaret C. Jacob. 1995. *Newton and the culture of Newtonianism.* Atlantic Highlands, New Jersey: Humanities Press.

Donovan, Arthur, Larry Laudan, and Rachel Laudan, eds. 1988. *Scrutinizing science: Empirical studies of scientific change.* Dordrecht: Kluwer.

Duhem, Pierre. 1962. *The aim and structure of physical theory.* Trans. Philip P. Wiener from *La théorie physique: son objet et structure* (2d ed., 1914). New York: Atheneum.

Dupré, John. 1993. *The disorder of things: Metaphysical foundations of the disunity of science.* Cambridge: Harvard University Press.

Durant, J. 1993. What is scientific literacy? In *Science and culture in Europe,* ed. J. Durant and J. Gregory. London: Science Museum, 136.

Dutton, D., and P. Henry. 1986. Truth matters. *Philosophy and Literature* 20: 299–304.

Einstein, Albert. 1905. Zur Elektrodynamik bewegter Körper. *Annalen der Physik und Chemie IV* 17:891–921.

———. 1929. *Festschrift für Aunel Stadola.* Zürich: Orell Füssli Verlag.

———. 1950. *Out of my later years.* New York: Philosophical Library.

———. 1954. *Ideas and opinions.* New York: Crown Publishers.

Evans, Lawrence. 1996. Should we care about science "studies"? *Duke Faculty Forum,* 1–3 October.

Feyerabend, Paul K. 1978. *Against method: Outlines of an anarchistic theory of knowledge.* 2d ed. London: Verso. Original edition, 1975.

Feynman, Richard. 1986. *Surely you're joking, Mr. Feynman.* London: Counterpoint.

Fine, Arthur. 1986. The natural ontological attitude. In *The shaky game: Einstein, realism and the quantum theory.* Chicago: University of Chicago Press, 112–35.

Fish, Stanley. 1996. Professor Sokal's bad joke. *New York Times,* 21 May, sec. A.

Fleck, Ludwik. 1979. *Genesis and development of a scientific fact.* Chicago: University of Chicago Press. Original edition, 1935.

Fodor, Jerry. 1998. Look! Review of *Consilience: The unity of knowledge,* by Edward O. Wilson. *London Review of Books* 20, no. 21 (29 October):3, 6.

Frank, Tom. 1996. Textual reckoning: A scholarly host puts a transgression-minded journal on the defensive. *In These Times,* 27 May, 22–24.

Franklin, A. 1994. How to avoid the experimenters' regress. *Studies in History and Philosophy of Science* 25, no. 3:463–91.

Friedman, Alan. 1995. Exhibits and expectations. *Public Understanding of Science* 4:306.

Friedman, Michael. 1998. On the sociology of scientific knowledge and its philosophical agenda. *Studies in History and Philosophy of Science* 29: 239–71.

Galison, Peter. 1987. *How experiments end.* Chicago: University of Chicago Press.

Gauntlett, David. 1995. *Moving experiences: Understanding television's influences and effects.* London: John Libbey.

Geison, Gerald L. 1995. *The private science of Louis Pasteur.* Princeton: Princeton University Press.

———. 1997. Letter to the editor. *New York Review of Books,* 4 April.

Gergen, Kenneth J. 1988. Feminist critique of science and the challenge of social epistemology. In *Feminist thought and the structure of knowledge,* ed. Mary McCanney Gergen. New York: New York University Press, 27–48.

Giddens, Anthony. 1990. *The consequences of modernity.* Cambridge, MA: Polity.

———. 1996. *Introduction to sociology,* 2d ed. New York: W. W. Norton.

Gieryn, Thomas F. 1999. *Cultural boundaries of science: Credibility on the line.* Chicago: University of Chicago Press.

Gilbert, G. Nigel, and Michael Mulkay. 1984. *Opening Pandora's box: An analysis of scientists' discourse.* Cambridge: Cambridge University Press.

Gingras, Yves. 1995. Un air de radicalisme: sur quelques tendances récentes en sociologie de la science et de la technologie. *Actes de la Recherche en Sciences Sociales* 108:3–17.

Goldstein, Sheldon. 1996. Quantum philosophy: The flight from reason in science. In *The flight from science and reason,* ed. Paul R. Gross, Norman Levitt, and Martin W. Lewis. New York: New York Academy of Sciences, 119–25.

Goodstein, David. 1996. Conduct and misconduct in science. In *The flight from science and reason,* ed. Paul R. Gross, Norman Levitt, and Martin W. Lewis. New York: New York Academy of Sciences, 31–38.

Gottfried, Kurt. 1997. Was Sokal's hoax justified? *Physics Today* 50, no. 1 (January):61–62.

Gottfried, Kurt, and Kenneth Wilson. 1997. Science as a cultural construct. *Nature* 386: 545–47.

Gregory, Jane, and Steve Miller. 1998. *Science in public: Communication, culture and credibility.* New York: Plenum.

Gross, Paul R. 1996. Introduction to *The flight from science and reason,* ed.

Paul R. Gross, Norman Levitt, and Martin W. Lewis. New York: New York Academy of Sciences, 1–7.

———. 1998a. Evidence-free forensics and enemies of objectivity. In *A house built on sand: Exposing postmodernist myths about science,* ed. Noretta Koertge. New York: Oxford University Press, 99–118.

———. 1998b. Letter to the editor. *Physics Today* 51, no. 4 (April):15.

Gross, Paul R., and Norman Levitt. 1994. *Higher superstition: The academic Left and its quarrels with science.* Baltimore: Johns Hopkins University Press.

Gross, Paul R., Norman Levitt, and Martin W. Lewis, eds. 1996. *The flight from science and reason.* New York: New York Academy of Sciences.

Haack, Susan. 1996. Towards a sober sociology of science. In *The flight from science and reason,* ed. Paul R. Gross, Norman Levitt, and Martin W. Lewis. New York: New York Academy of Sciences, 259–66.

———. 1997. We pragmatists . . . : Peirce and Rorty in conversation. *Partisan Review* 64, no. 1: 91–107. Reprinted in Haack 1998, chapter 2.

———. 1998. *Manifesto of a passionate moderate: Unfashionable essays.* Chicago: University of Chicago Press.

Hacking, Ian. 1997. Taking bad arguments seriously: Ian Hacking on psychopathology and social construction. *London Review of Books,* 21 August, 14–16.

———. 1999. *The social construction of what?* Cambridge: Harvard University Press.

Hacohen, Malachi H. 1998. Karl Popper, the Vienna Circle, and Red Vienna. *Journal of the History of Ideas* 59:711–34.

Hanson, Norwood Russell. 1965. *Patterns of discovery: An inquiry into the conceptual foundations of science.* Cambridge: Cambridge University Press.

Harrison, Edward. 1987. Whigs, Prigs, and Historians of Science. *Nature* 329 (17 September):213–14.

Hart, Roger. 1996. The flight from reason: Higher superstition and the refutation of science studies. In *Science wars,* ed. Andrew Ross. Durham, NC: Duke University Press, 259–92.

Hellman, Hal. 1998. *Great feuds in science: Ten of the liveliest disputes ever.* New York: John Wiley and Sons, 1998.

Hesse, Mary B. 1980. *Revolutions and reconstructions in the philosophy of science.* Brighton: Harvester.

Hodges, Andrew. 1983. *Alan Turing: The enigma of intelligence.* London: Counterpoint.

Holton, Gerald. 1992. How to think about the "anti-science" phenomenon. *Public Understanding of Science* 1:103–28.

———. 1993. *Science and anti-science.* Cambridge: Harvard University Press.

———. 1996. *Einstein, history, and other passions.* Reading, MA.: Addison-Wesley.

Horgan, John. 1996. *The end of science: Facing the limits of knowledge in the twilight of the scientific age.* Reading, MA: Helix Books.

Horgan, John, and John Maddox. 1998. Resolved: Science is at an end. Or is it? *New York Times,* 10 November, sec. D.

Hornig, Susanna. 1993. Reading risk: Public response to accounts of technological risk. *Public Understanding of Science* 2:98.

Hoyningen-Huene, Paul. 1993. *Reconstructing scientific revolutions: Thomas S. Kuhn's philosophy of science.* Chicago: University of Chicago Press.

Hubbard, Ruth. 1996. Gender and genitals: Constructs of sex and gender. In *Science wars,* ed. Andrew Ross. Durham, NC: Duke University Press, 168–79.

Hume, David. 1988. *An enquiry concerning human understanding.* 1748. Reprint, Amherst, NY: Prometheus.

Huxley, Thomas Henry. 1900. On the educational value of the natural history sciences. 1854. Reprint, in *Science and Education: Essays.* Vol. 3 of *Collected Essays,* New York: D. Appleton, 38–65.

Jardine, Nick, and Marina Frasca-Spada. 1997. Splendours and miseries of the science wars. *Studies in History and Philosophy of Science* 28:219–35.

Kapitza, Sergei. 1991. Anti-science trends in the USSR. *Scientific American,* August, 18–24.

Kinoshita, Toichiro. 1995. New value of the α^3 electron anomalous magnetic moment. *Physical Review Letters* 75:4728–31.

Kitcher, Phillip. 1998. A plea for science studies. In *A house built on sand: Exposing postmodernist myths about science,* ed. Noretta Koertge. New York: Oxford University Press, 32–56.

Koertge, Noretta, ed. 1998. *A house built on sand: Exposing postmodernist myths about science.* New York: Oxford University Press.

———. 1998a. Postmodernisms and the problem of scientific literacy. In *A house built on sand: Exposing postmodernist myths about science,* ed. Noretta Koertge. New York: Oxford University Press, 257–71.

Kuhn, Thomas S. 1977. The essential tension: Tradition and innovation in scientific research. In *The essential tension: Selected studies in scientific tradition and change.* Chicago: University of Chicago Press, 225–39.

———. 1992. The trouble with the historical philosophy of science. Rothschild Distinguished Lecture, 19 November 1991, Harvard University, Department of the History of Science, Cambridge.

———. 1996. *The structure of scientific revolutions.* 3d ed. Chicago: University of Chicago Press. Original edition, 1962.

Kurtz, Paul. 1996. Two sources of unreason in democratic society: The paranormal and religion. In *The flight from science and reason,* ed. Paul R. Gross, Norman Levitt, and Martin Lewis. New York: New York Academy of Sciences, 493–504.

Labinger, Jay A. 1995. Science as culture: A view from the petri dish. *Social Studies of Science* 25:285–306.

———. 1997. The science wars and the future of the American academic profession. *Daedalus* 126, no. 4 (fall):201–20.

Lakatos, Imre. 1970. Falsification and the methodology of scientific research programmes. In *Criticism and the growth of knowledge,* ed. Imre Lakatos and Alan Musgrave. Cambridge: Cambridge University Press, 91–196.

———. 1978. *The methodology of scientific research programmes.* Cambridge: Cambridge University Press.

Latour, Bruno. 1987. *Science in action: How to follow scientists and engineers through society.* Cambridge: Harvard University Press.

———. 1990. Postmodern? No, simply *a*modern. Steps towards an anthropology of science: An essay review. *Studies in History and Philosophy of Science* 21:145–71.

———. 1998. Ramsès II est-il mort de la tuberculose? *La Recherche* 307 (March):84–85; errata, 308 (April):85 and 309 (May):7.

———. 1999a. For David Bloor . . . and beyond: A reply to David Bloor's "anti-Latour." *Studies in History and Philosophy of Science* 30:113–29.

———. 1999b. *Pandora's hope: Essays on the reality of science studies.* Cambridge: Harvard University Press.

Latour, Bruno, and Steve Woolgar. 1986. *Laboratory life: The construction of scientific facts.* 2d ed. Princeton: Princeton University Press.

Laubscher, Roy Edward. 1981. *Astronomical papers prepared for the use of the American Ephemeris and Nautical Almanac* 22, parts 2 and 4. Washington, DC: US Government Printing Office.

Laudan, Larry. 1981. The pseudo-science of science? *Philosophy of the Social Sciences* 11:173–98.

———. 1990a. *Science and relativism.* Chicago: University of Chicago Press.

———. 1990b. Demystifying underdetermination. *Minnesota Studies in the Philosophy of Science* 14:267–97.

Levitt, Norman. 1996. Mathematics as the stepchild of contemporary culture. In *The flight from science and reason,* ed. Paul R. Gross, Norman Levitt, and Martin Lewis. New York: New York Academy of Sciences, 39–53.

Lewenstein, Bruce V. 1996. Shooting the messenger: Understanding attacks on science in American life. Paper presented at the Fourth International Conference on Public Communication of Science and Technology, 24 November, Melbourne, Australia.

Lewontin, Richard C. 1993. *Biology as ideology: The doctrine of DNA.* New York: HarperPerennial.

———. 1996. A la recherche du temps perdu: A review essay. In *Science wars,* ed. Andrew Ross. Durham, NC: Duke University Press, 293–301.

———. 1998. Survival of the nicest? *New York Review of Books,* 22 October, 59–63.

Lynch, Michael. 1988. Sacrifice and the transformation of the animal body into a scientific object: Laboratory culture and ritual practice in the neurosciences. *Social Studies of Science* 18:265–89.

———. 1993. *Scientific practice and ordinary action.* New York: Cambridge University Press.

———. 1997. A so-called "fraud": Moral modulations in a literary scandal. *History of the Human Sciences* 10, no. 3:9–21.

Maddox, John, James Randi, and Walter W. Stewart. 1988. "High-dilution" experiments a delusion. *Nature* 334:287–90.

———. 1998. *What remains to be discovered?* New York: Free Press.

Mahoney, Michael J. 1979. Psychology of the scientist: An evaluative review. *Social Studies of Science* 9:349–75.

Mahoney, Michael J., and B. G. DeMonbreun. 1977. Psychology of the scientist: An analysis of problem-solving bias. *Cognitive Therapy and Research* 1:229–38.

Martin, Brian, Evelleen Richards, and Pam Scott. 1991. Who's a captive? Who's a victim? Response to Collins's method talk. *Science, Technology, and Human Values* 16:252–55.

Mayr, Ernst. 1997. *This is biology*. Cambridge: Harvard University Press.

McGinn, Colin. 1993. *Problems in philosophy: The limits of inquiry*. Oxford: Blackwell.

McKinney, William J. 1998. When experiments fail: Is "cold fusion" science as normal? In *A house built on sand: Exposing postmodernist myths about science,* ed. Noretta Koertge. New York: Oxford University Press, 133–50.

Mermin, N. David. 1996a. What's wrong with this sustaining myth? *Physics Today* 49, no. 3 (March):11–13.

———. 1996b. The golemization of relativity. *Physics Today* 49, no. 4 (April): 11–13.

———. 1996c. Sociologists, scientist continue debate about scientific process. *Physics Today* 49, no. 7 (July):11–15, 88.

———. 1997. Sociologists, scientist pick at threads of argument about science. *Physics Today* 50, no. 1 (January):92–95.

———. 1998a. The science of science: A physicist reads Barnes, Bloor, and Henry. *Social Studies of Science* 28:603–23.

———. 1998b. Abandoning preconceptions: Reply to Bloor and Barnes. *Social Studies of Science* 28:641–47.

———. 1999. Border control at the frontiers of science. *Nature* 401:328.

Miller, Jon D. 1987. Scientific literacy in the United States. In *Communicating science to the public,* ed. D. Evered and M. O'Connor. New York: Wiley, 14–19.

Mirowski, Philip. 1994. A visible hand in the marketplace of ideas: Precision measurement as arbitrage. *Science in Context* 7, no. 3:563–89.

Monk, Ray. 1990. *Ludwig Wittgenstein, the duty of genius.* New York: Free Press.

Mulkay, Michael J., and G. Nigel Gilbert. 1981. Putting philosophy to work: Karl Popper's influence on scientific practice. *Philosophy of the Social Sciences* 11:389–407.

Mullis, Kary. 1998. *Dancing naked in the mind field.* New York: Pantheon Books.

Nagel, Thomas. 1997. *The last word.* New York: Oxford University Press.

Nanda, Meera. 1997. The science wars in India. *Dissent* 44, no. 1 (winter): 78–83.

Neidhardt, F. 1993. The public as a communication system. *Public Understanding of Science* 2:339–50.

Nelkin, Dorothy. 1995. *Selling science: How the press covers science and technology.* New York: W. H. Freeman.

Norton, Mary Beth, and Pamela Gerardi, eds. 1995. *The American Historical Association's guide to historical literature.* 3d ed. Vols. 1 and 2. New York: Oxford University Press.

Nowotny, Helga. 1979. Science and its critics: Reflections on anti-science. In *Counter-movements in the sciences,* ed. H. Nowotny and H. Rose. Dordrecht: Reidel.

Oppenheimer, J. Robert. 1954. *Science and the common understanding: The BBC Reith Lectures 1953*. New York: Oxford University Press.

———. 1955. The scientist in society. In *The open mind*. New York: Simon and Schuster, 119–29.

Park, Robert. 1994. Is science the god that failed? *Science Communication* 16, no. 2:207.

Parkin, Gerard. 1992. Do bond-stretch isomers really exist? *Accounts of Chemical Research* 25:455–60.

Perutz, M. F. 1995. The pioneer defended. *New York Review of Books,* 21 December.

———. 1997. Letter to the editor. *New York Review of Books,* 4 April.

Petley, Brian. 1985. *The fundamental physical constants and the frontiers of measurement.* Bristol: Adam Hilger.

Pickering, Andrew. 1984. *Constructing quarks*. Chicago: University of Chicago Press.

———. 1987. Forms of life: Science, contingency and Harry Collins. *British Journal for the History of Science* 20:213–21.

———. 1992. *Science as practice and culture.* Chicago: University of Chicago Press.

———. 1995. *The mangle of practice: Time, agency, and science.* Chicago: University of Chicago Press.

———. 1998. Review of *Image and Logic,* by P. Galison. *Times Literary Supplement,* 24 July.

Pinch, Trevor. 1984. Relativism—is it worth the candle? Paper presented to the History of Science Society, October 12–16, New Orleans.

———. 1986. *Confronting nature: The sociology of solar-neutrino detection.* Dordrecht: Kluwer.

———. 1995. Rhetoric and the cold fusion controversy: From the chemists' Woodstock to the physicists' Altamont. In *Science, reason, and rhetoric,* ed. Henry Krips, J. E. McGuire, and Trevor Melia. Pittsburgh: University of Pittsburgh Press, 153–76.

———. 1999. Half a house: A response to McKinney. *Social Studies of Science* 29, no. 2:235–40.

Pinnick, Cassandra L. 1998. What is wrong with the Strong Programme's case study of the "Hobbes-Boyle Dispute"? In *A house built on sand: Exposing postmodernist myths about science,* ed. Noretta Koertge. New York: Oxford University Press, 227–39.

Planck, Max. 1949. *Scientific autobiography and other papers.* Trans. Frank Gaynor. New York: Philosophical Library.

Pleat, F. David. 1997. *Infinite potential: The life and times of David Bohm.* Reading, MA: Helix Books.

Plotnitsky, Arkady. 1997. But it is above all not true: Derrida, relativity, and the "science wars." *Postmodern Culture* 7, no. 2:1–27.

Polanyi, Michael. 1967. *The tacit dimension.* New York: Anchor.

Popper, Karl R. 1957. *The poverty of historicism.* London: Routledge and Kegan Paul.

———. 1959. *The logic of scientific discovery.* London: Hutchinson.

———. 1972. Science: Conjectures and refutations. In *Conjectures and refuta-*

tions: The growth of scientific knowledge. 4th, rev. ed. London: Routledge and Kegan Paul, 33–65.

———. 1976. *Unended quest: An intellectual autobiography.* London: Fontana/ Collins.

Quine, Willard Van Orman. 1980. Two dogmas of empiricism. In *From a logical point of view.* 2d, rev. ed. Cambridge: Harvard University Press. Original edition, 1953.

Rabinow, Paul. 1996. *Making PCR: A story of biotechnology.* Chicago: University of Chicago Press.

Radder, Hans. 1998. The Politics of STS. *Social Studies of Science* 28:325–31.

Roll-Hansen, Nils. 2000. The application of complementarity to biology: From Niels Bohr to Max Delbrück. *Historical Studies in the Physical and Biological Sciences* 30, no. 2:417–42.

Rorty, Richard. 1997. Thomas Kuhn, rocks, and the laws of physics. *Common Knowledge* 6, no. 1:6–16.

———. 1998. *Truth and progress: Philosophical papers.* Cambridge: Cambridge University Press.

Ross, Andrew, ed. 1996. *Science wars.* Durham, NC: Duke University Press.

———. 1996b. Introduction to *Science wars,* ed. Andrew Ross. Durham, NC: Duke University Press, 1–15.

Rudwick, Martin. 1985. *The great Devonian controversy.* Chicago: University of Chicago Press.

Russell, Bertrand. 1949. *The practice and theory of Bolshevism.* 2d ed. London: George Allen and Unwin. Original edition, 1920.

———. 1961. *History of Western Philosophy.* 2d ed. London: George Allen and Unwin. Original edition, 1946.

Sampson, Wallace. 1996. In *The flight from science and reason,* ed. Paul R. Gross, Norman Levitt, and Martin Lewis. New York: New York Academy of Sciences, 188–97.

Schaffer, Simon. 1991. Utopia unlimited: On the end of science. *Strategies* 4/5:151–81.

Scott, Pam, Evelleen Richards, and Brian Martin. 1990. Captives of controversy: The myth of the neutral social researcher in contemporary scientific controversies. *Science, Technology, and Human Values* 15:474–94.

Shapin, Steven. 1989. The invisible technician. *American Scientist* 77:554–63.

———. 1994. *A social history of truth: Civility and science in seventeenth-century England.* Chicago: University of Chicago Press.

———. 1995a. Here and everywhere: Sociology of scientific knowledge. *Annual Review of Sociology* 21:289–321.

———. 1995b. Cordelia's love: Credibility and the social studies of science. *Perspectives on Science* 3:255–75.

———. 1999. Rarely pure and never simple: Talking about truth. *Configurations* 7:1–14.

———. Forthcoming. Truth and credibility: Science and the social study of science. In *International encyclopedia of social and behavioral sciences,* Neil J. Smelser and Paul B. Baltes, gen. eds., Sheila Jasanoff, sec. ed. for science and technology studies. Oxford: Elsevier Science.

Shapin, Steven, and Simon Schaffer. 1985. *Leviathan and the air-pump:*

Hobbes, Boyle, and the experimental life. Princeton: Princeton University Press.

Sharrock, Wes, and Bob Anderson. 1991. Epistemology: Professional scepticism. In *Ethnomethodology and the Human Sciences,* ed. Graham Button. Cambridge: Cambridge University Press, 51–76.

Shulman, Robert G. 1998. Hard days in the trenches. *FASEB Journal* 12:255–58.

Slezak, Peter. 1994a. A second look at David Bloor's *Knowledge and social imagery. Philosophy of the Social Sciences* 24:336–61.

———. 1994b. The social construction of social constructionism. *Inquiry* 37: 139–57.

Smith, George E. Forthcoming. From the phenomenon of the ellipse to an inverse-square force: Why not? In *Festschrift in honor of Howard Stein's seventieth birthday,* ed. David Malament. La Salle, Illinois: Open Court.

Snow, C. P. 1959. *The two cultures and the scientific revolution.* New York: Cambridge University Press.

Sokal, Alan. 1996a. A physicist experiments with cultural studies. *Lingua Franca,* May/June, 62–64.

———. 1996b. Transgressing the boundaries: Toward a transformative hermeneutics of quantum gravity. *Social Text* 46/47:217–52.

———. 1996c. Truth or consequences: A brief response to Robbins. *Tikkun,* November/December, 58.

———. 1998. What the social text affair does and does not prove. In *A house built on sand: Exposing postmodernist myths about science,* ed. Noretta Koertge. New York: Oxford University Press, 9–22.

Sokal, Alan, and Jean Bricmont. 1998. *Intellectual impostures: Postmodern philosophers' abuse of science.* London: Profile Books. Published in the US and Canada under the title *Fashionable nonsense: Postmodern intellectuals' abuse of science.* New York: Picador USA. Originally published in French under the title *Impostures intellectuelles.* Paris: Odile Jacob, 1997. 2d ed. Paris: Livre de Poche, 1999.

Standish, Myles E., Jr. 1993. Planet X: No dynamical evidence in the optical observations. *Astronomical Journal* 105:2000.

Stent, Gunther. 1969. *The coming of the golden age: A view of the end of progress.* Garden City, NY: Natural History Press.

Suchman, Lucy. 1987. *Plans and situated actions.* Cambridge: Cambridge University Press.

Sullivan, Philip A. 1998. An engineer dissects two case studies: Hayles on fluid mechanics and Mackenzie on statistics. In *A house built on sand: Exposing postmodernist myths about science,* ed. Noretta Koertge. New York: Oxford University Press, 71–98.

Sulloway, Frank. 1996. *Born to rebel: Birth order, family dynamics, and creative lives.* New York: Vintage Books.

Summers, William C. 1997. Letter to the editor. *New York Review of Books,* 6 February.

Trachtman, Leon E. 1981. The public understanding of science effort: A critique. *Science, Technology, and Human Values* 6, no. 3:10–12.

Urbach, Peter. 1987. *Francis Bacon's philosophy of science.* La Salle, IL: Open Court.

Van Dyck, Robert S., Jr., Paul B. Schwinberg, and Hans G. Dehmelt. 1987. New high-precision comparison of electron and positron *g* factors. *Physical Review Letters* 59:26–29.

Weinberg, Steven. 1992. *Dreams of a final theory.* New York: Pantheon.

———. 1995. Night thoughts of a quantum physicist. *Bulletin of the American Academy of Arts and Sciences* 49:51–64.

———. 1996. Sokal's Hoax. *New York Review of Books* 43, no. 13 (August 8): 11–15.

———. 1998. The revolution that didn't happen. *New York Review of Books* 45, no. 15 (8 October):48–52.

Wilson, Curtis A. 1980. Perturbations and solar tables from Lacaille to Delambre: The rapproachment of observation and theory, part 1. *Archive for History of Exact Sciences* 22:54.

Wilson, E. Bright. 1990. *An introduction to scientific research.* New York: Dover.

Wilson, Edward O. 1995. *Naturalist.* New York: Warner Books.

———. 1998a. Back from chaos. *Atlantic Monthly,* March, 41–62.

———. 1998b. *Consilience: The unity of knowledge.* New York: Knopf.

Winch, Peter. 1958. *The idea of a social science and its relation to philosophy.* London: Routledge and Kegan Paul.

Wittgenstein, Ludwig. 1953. *Philosophical investigations.* Oxford: Blackwell.

Wolpert, Lewis. 1992. *The unnatural nature of science: Why science does not make (common) sense.* London: Faber and Faber.

Wynne, Brian. 1996. Misunderstood misunderstandings: Social identities and public uptake of science. In *Misunderstanding science? The public reconstruction of science and technology,* ed. A. Irwin and B. Wynne. Cambridge: Cambridge University Press, 19–46.

Contributors

Constance Barsky is a geochemist who has been involved in education reform initiatives since 1992. Her prior research on Thomas Kuhn and the history of scientific revolutions and her experience in industrial research and development have influenced her approach to examining issues surrounding education reform. She is coauthor and coeditor of the fall 1998 issue of *Daedalus,* entitled "Education Yesterday, Education Tomorrow." Her current research focuses on historical processes of continuing improvement in science and technology, current processes of strategic thinking, and their potential applications to education.

Constance K. Barsky
Learning by Redesign
Department of Physics
Smith Laboratory
The Ohio State University
174 West 18th Avenue
Columbus, OH 43210
USA
barsky.1@osu.edu

Jean Bricmont is Professor of Theoretical Physics at the University of Louvain, Belgium. He is coauthor (with Alan Sokal) of *Fashionable Nonsense: Postmodern Intellectuals' Abuse of Science* (1998). He works in the fields of statistical physics and partial differential equations and also on the foundations of statistical physics and quantum mechanics. He is author of "Science of Chaos, or Chaos in Science?" a critique of various popularizations of science, especially those due to Prigogine.

Jean Bricmont
FYMA, 2, ch. du Cyclotron, UCL
B-1348 Louvain-la-neuve
Belgium
bricmont@fyma.ucl.ac.be

Harry Collins is Distinguished Research Professor of Sociology and director of the Centre for the Study of Knowledge, Expertise, and Science (KES) at Cardiff University. He is the author of around one hundred papers and half a dozen monographs on topics to do with the social nature of scientific knowledge and the difficulty of mimicking social knowledge with intelligent machines. His most well known books are *Changing Or-*

Harry Collins
KES
Cardiff School of Social Sciences
Cardiff University
The Glamorgan Building
King Edward VIIth Avenue
Cardiff CF10 3WT
UK
CollinsHM@cardiff.ac.uk
www.cardiff.ac.uk/socsi/whoswho.htm

der: Replication and Induction in Scientific Practice; Artificial Experts: Social Knowledge and Intelligent Machines; and *The Golem* series, written with Trevor Pinch. With Martin Kusch he has recently published *The Shape of Actions: What Humans and Machines Can Do.*

Peter Dear specializes in the history of early-modern science. He is author of *Discipline and Experience: The Mathematical Way in the Scientific Revolution* (1995) and, most recently, *Revolutionizing the Sciences: European Knowledge and Its Ambitions, 1500–1700* (forthcoming). He is currently working on a book tentatively entitled *Making Sense in Science.*

Peter Dear
Department of Science and Technology Studies
632 Clark Hall
Cornell University
Ithaca, NY 14853
USA
prd3@cornell.edu

Jane Gregory is Lecturer in Science and Technology Studies and Course Director for Science and Society at Birkbeck College, University of London. She is coauthor with Steve Miller of *Science in Public: Communication, Culture, and Credibility* (1998). Her research interests include communication, popularization, and unorthodox science.

Jane Gregory
Birkbeck College
University of London
26 Russell Square
London WC1B 5DQ
UK
j.gregory@bbk.ac.uk

Jay Labinger is a chemist by training and continues to be active in research in catalysis and organometallic chemistry. In the last eight years or so he has also become involved in areas encompassing broader issues of science and has written articles and spoken at meetings on science and literature, sociology of science, and history of science. His current position is administrator of the Beckman Institute at Caltech.

Jay Labinger
California Institute of Technology
139-74
Pasadena, CA 91125
USA
jal@cco.caltech.edu

Michael Lynch has a background in the sociological subfield of ethnomethodology (the study of practical action and practical reasoning in everyday and professional circumstances). He uses that approach to investigate informal "shop talk" among scientists and to examine the production of measurements and

Michael Lynch
Department of Science and Technology Studies
632 Clark Hall
Cornell University
Ithaca, NY 14853
USA
mel27@cornell.edu

visual representations in day-to-day laboratory work. He has also studied the organization of courtroom testimony and is currently studying the intersection of law and science in criminal investigations involving DNA profiling. His book, *Scientific Practice and Ordinary Action* (1993), reviews and critically examines work in social studies of science.

N. David Mermin
Department of Physics
109 Clark Hall
Cornell University
Ithaca, NY 14853
USA
ndm4@cornell.edu

David Mermin is a theoretical physicist. Over the years he has contributed to solid state physics, statistical physics, low temperature physics, crystallography, and foundations of quantum mechanics. His most recent research interests are in the new field of quantum computation. He is coauthor of the standard introductory text *Solid State Physics* (1976). Since 1988 he has contributed about two dozen nontechnical columns about physics and the practice of physics to *Physics Today*. Many of his nontechnical writings are collected in *Boojums All the Way Through* (1990).

Steve Miller
Science and Technology Studies/
Physics and Astronomy
University College London
Gower Street
London WC1E 6BT
UK
s.miller@ucl.ac.uk

Steve Miller is Reader in Science Communication and Planetary Science at University College London and a former political journalist. His research interests in science and technology studies include the popularization of the physical sciences and the European dimension of public understanding of science issues and policy. With Jane Gregory, he is coauthor of *Science in Public: Communication, Culture, and Credibility* (1998). In planetary astronomy, his main interests are the giant planets, particularly the physics of the upper atmosphere, and the study of newly detected extrasolar planets; he has coauthored over eighty peer-reviewed articles in molecular physics and planetary science.

Trevor Pinch does research in sociology of science and technology. He is coauthor with Harry Collins of *The Golem: What You Should Know about Science* (1998; 2d edition) and *The Golem at Large: What You Should Know about Technology* (1998). His latest book, with Frank Trocco, is *Analog Days: The Invention and Impact of the Moog Synthesizer* (forthcoming).

Trevor Pinch
Department of Science and Technology Studies
632 Clark Hall
Cornell University
Ithaca, NY 14853
USA
tjp2@cornell.edu

Peter Saulson is a physicist who has worked on gravitational wave detection since 1981. He is author of *Fundamentals of Interferometric Gravitational Wave Detectors* (1994). On the faculty at Syracuse University since 1991, he spent 2000 helping to commission the LIGO interferometer at Livingston, Louisiana.

Peter R. Saulson
Department of Physics
Syracuse University
Syracuse, NY 13244
USA
saulson@phy.syr.edu

Steven Shapin's recent publications include: *A Social History of Truth* (1994); *The Scientific Revolution* (1996); and (coedited with Christopher Lawrence) *Science Incarnate: Historical Embodiments of Natural Knowledge* (1998). He is currently working on a cultural history of the scientist's role.

Steven Shapin
Department of Sociology, 0533
University of California, San Diego
La Jolla, CA 92093
USA
sshapin@ucsd.edu

Alan Sokal is Professor of Physics at New York University. His main research interests are in mathematical physics, more particularly in the theory of phase transitions. He is coauthor with Jürg Fröhlich and Roberto Fernández of *Random Walks, Critical Phenomena, and Triviality in Quantum Field Theory* (1992) and with Jean Bricmont of *Fashionable Nonsense: Postmodern Intellectuals' Abuse of Science* (1998).

Alan Sokal
Department of Physics
New York University
4 Washington Place
New York, NY 10003
USA
sokal@nyu.edu

Steven Weinberg is Professor of Physics and Astronomy at the University of Texas, Austin. His research has been in cosmology and in the theory of elementary particles, for which he was awarded the 1979 Nobel Prize in physics. His latest book for the general reader is *Dreams of a Final Theory* (1992).

Steven Weinberg
Department of Physics
University of Texas
Austin, TX 78712
USA
weinberg@physics.utexas.edu

Kenneth Wilson, winner of the 1982 Nobel Prize in physics, is author of many articles in elementary particle physics and statistical mechanics. He turned his interests to education reform and related social science research ten years ago. He is coauthor of the book *Redesigning Education* and coauthor and coeditor for the fall 1998 issue of *Daedalus,* entitled "Education Yesterday, Education Tomorrow." His research focus is on future directions for education and their possible relationships to past processes of continuing improvement in science, technology, and education.

Kenneth G. Wilson
Learning by Redesign
Department of Physics
Smith Laboratory
The Ohio State University
174 West 18th Avenue
Columbus, OH 43210
USA
kgw@pacific.mps.ohio-state.edu

Index